A GUIDE TO THE ELEMENTS

THIRD EDITION

A GUIDE TO THE ELEMENTS

ALBERT STWERTKA

OXFORD
UNIVERSITY PRESS

OXFORD
UNIVERSITY PRESS

Oxford University Press, Inc., publishes works that further Oxford University's objective of excellence in research, scholarship, and education.

Oxford New York
Auckland Cape Town Dar es Salaam Hong Kong Karachi
Kuala Lumpur Madrid Melbourne Mexico City Nairobi
New Delhi Shanghai Taipei Toronto

With offices in
Argentina Austria Brazil Chile Czech Republic France Greece
Guatemala Hungary Italy Japan Poland Portugal Singapore
South Korea Switzerland Thailand Turkey Ukraine Vietnam

Published by Oxford University Press, Inc.
198 Madison Avenue, New York, New York 10016
www.oup.com

Consultant: Robin Eichen Conn, Cargill, Inc.
Design: Valerie Sauers, Patrice Sheridan
Picture research: Amla Sanghvi, Jennifer Traslavina, Sonia Tycko, Hanna Oldsman

Library of Congress Cataloging-in-Publication Data
Stwertka, Albert.
A guide to the elements / Albert Stwertka.—3rd ed.
p. cm.
ISBN 978-0-19-983252-1 (pbk.)—ISBN 978-0-19-983251-4 (hardcover)
1. Chemical elements. 2. Periodic law. I. Title.
QD466.S78 2012
546'.8—dc23
2011020804

ISBN 9780199832521 (paperback)
9780199832514 (hardcover)

1 3 5 7 9 8 6 4 2

Printed in the United States of America
on acid-free paper

Contents

Periodic Table of the Elements

Key: 1 — Atomic number; H — Chemical symbol; 1.00794 — Atomic weight

Periods	1 IA	2 IA	3 IIIB	4 IVB	5 VB	6 VIB	7 VIIB	8 VIIIB	9 VIIIB	10 VIIIB	11 IB	12 IIB	13 IIIA	14 IVA	15 VA	16 VIA	17 VIIA	18 VIIIA
1	1 H 1.00794																	2 He 4.00260
2	3 Li 6.941	4 Be 9.01218											5 B 10.811	6 C 12.011	7 N 14.00674	8 O 15.9994	9 F 18.99840	10 Ne 20.1797
3	11 Na 22.98977	12 Mg 24.3050											13 Al 26.98154	14 Si 28.0855	15 P 30.97376	16 S 32.066	17 Cl 35.4527	18 Ar 39.948
4	19 K 39.0983	20 Ca 40.078	21 Sc 44.95591	22 Ti 47.88	23 V 50.9415	24 Cr 51.9961	25 Mn 54.9380	26 Fe 55.847	27 Co 58.93320	28 Ni 58.6934	29 Cu 63.546	30 Zn 65.39	31 Ga 69.723	32 Ge 72.61	33 As 74.92159	34 Se 78.96	35 Br 79.904	36 Kr 83.80
5	37 Rb 85.4678	38 Sr 87.62	39 Y 88.90585	40 Zr 91.224	41 Nb 92.90638	42 Mo 95.94	43 Tc** 98.9072	44 Ru 101.07	45 Rh 102.90550	46 Pd 106.42	47 Ag 107.8682	48 Cd 112.411	49 In 114.82	50 Sn 118.710	51 Sb 121.76	52 Te 127.60	53 I 126.09447	54 Xe 131.29
6	55 Cs 132.905	56 Ba 137.327	57 *La 138.9055	72 Hf 178.49	73 Ta 180.9479	74 W 183.85	75 Re 186.207	76 Os 190.2	77 Ir 192.22	78 Pt 195.08	79 Au 196.96654	80 Hg 200.59	81 Tl 204.3833	82 Pb 207.2	83 Bi 208.98037	84 Po** 208.9824	85 At** 209.9871	86 Rn** 222.0176
7	87 Fr** 223.0197	88 Ra** 226.0254	89 †Ac** 227.0278	104 Rf** 261.11	105 Db** 262.114	106 Sg** 263.118	107 Bh** 262.12	108 Hs** (265)	109 Mt** (266)	110 Ds** (281)	111 Rg** (281)	112 Cn** (285)	113 Uut** (284)	114 Uuq** (289)	115 Uup** (288)	116 Uuh** (293)	117 Uus** (294)	118 Uuo** (294)

*	58 Ce 140.115	59 Pr 140.90765	60 Nd 144.24	61 Pm** 144.9127	62 Sm 150.36	63 Eu 151.965	64 Gd 157.25	65 Tb 158.92534	66 Dy 162.50	67 Ho 164.93032	68 Er 167.26	69 Tm 168.93421	70 Yb 173.04	71 Lu 174.967
†	90 Th** 232.0381	91 Pa** 231.0359	92 U** 238.0289	93 Np** 237.0482	94 Pu** 244.0642	95 Am** 243.0614	96 Cm** 247.0703	97 Bk** 247.0703	98 Cf** 251.0796	99 Es** 252.083	100 Fm** 257.0951	101 Md** 258.10	102 No** 259.1009	103 Lr** 262.11

**All the isotopes of this element are radioactive. With the exception of uranium and thorium, the atomic weight shown represents the relative atomic weight of the longest-lived isotope. The numbers in parentheses are the mass numbers of the longest-lived isotope.

* Asterisk - lanthanides; † Dagger - actinides

The Periodic Table

Although the world is varied and complex, everything in it—air, water, rocks, living tissue, and the almost infinite number of other objects and materials around us—is actually made up of only a limited number of chemical elements. We know today that only 91 such elements exist naturally on the Earth. They range from hydrogen, the lightest element, to uranium, the heaviest. Actually, several more elements do exist, but these have to be made artificially in laboratories.

The basic components of each chemical element are atoms. The atoms of an element consist of three kinds of particles: protons, neutrons, and electrons. Protons and neutrons exist at the core, or nucleus, of the atom. One of the important ways in which these two kinds of particles differ from each other is that each proton carries a single, positive electric charge, whereas a neutron carries no electric charge. Electrons, which are much smaller than either protons or neutrons, each carry a single negative electric charge. Electrons are present at some distance away from the nucleus of the atom and travel rapidly around it in complex paths known as orbits. Under normal circumstances, the number of electrons orbiting around the nucleus of a particular atom is exactly equal to the number of protons in the nucleus of the atom, so that the overall positive electric charge provided by its protons is exactly balanced by the overall negative charge provided by the electrons orbiting its nucleus.

The unique properties of each of the chemical elements are determined by their number of neutrons, protons, and electrons. Besides determining the properties of a pure chemical element, the neutron, proton, and electron content of its atoms also determines its behavior in relation to other chemical elements. Although each element behaves differently and has different properties from all of the others, the atoms of different elements can combine with one another to form clusters of atoms called molecules. It is this combination of atoms that accounts for the enormous variety of chemical substances that can be found in nature and created by modern technology.

When scientists first tried to describe the physical and chemical properties of the elements and chemical compounds, which are formed by the combination of atoms of different elements, they soon became buried under a mountain of seemingly unconnected facts. Many early scientists recognized the need to organize this information, and they attempted to discover some sort of order or pattern that could simplify what seemed to them an overwhelming array of chemical facts. The solution to the

The Russian chemist Dmitry Mendeleyev created his periodic table (below) in 1869 while preparing a chemistry textbook for his students.

problem was the so-called periodic table of the chemical elements.

The modern periodic table is based primarily on the work of the Russian chemist Dmitry Ivanovich Mendeleyev (1834–1907) and the German physicist Julius Lothar Meyer (1830–1895). Working independently, these scientists developed similar periodic tables within a few months of each other in 1869. Mendeleyev, however, is usually given the credit for having developed the periodic table because he managed to publish his work first.

Mendeleyev, who was a professor of chemistry at the University of St. Petersburg, developed the periodic table while preparing a chemistry textbook for his students. As part of this project, he had written down the properties of the elements on cards. While sorting through these cards he noticed that when the elements were arranged in order of their weight, similar chemical properties occurred repeatedly at regular intervals. Mendeleyev used this observation to construct his periodic table. He placed the elements in horizontal rows, according to their weight, with the lightest in each row at the left end of the row and the heaviest at the right, and with one row beneath the other, so that all elements with similar properties fell into vertical columns.

xviii PRINCIPLES OF CHEMISTRY

PERIODIC SYSTEM OF THE ELEMENTS IN GROUPS AND SERIES.

Series	GROUPS OF ELEMENTS								
	0	I	II	III	IV	V	VI	VII	VIII
1	—	Hydrogen H 1·008	—	—	—	—	—	—	
2	Helium He 4·0	Lithium Li 7·03	Beryllium Be 9·1	Boron B 11·0	Carbon C 12·0	Nitrogen N 14·04	Oxygen O 16·00	Fluorine F 19·0	
3	Neon Ne 19·9	Sodium Na 23·05	Magnesium Mg 24·3	Aluminium Al 27·0	Silicon Si 28·4	Phosphorus P 31·0	Sulphur S 32·06	Chlorine Cl 35·45	
4	Argon Ar 38	Potassium K 39·1	Calcium Ca 40·1	Scandium Sc 44·1	Titanium Ti 48·1	Vanadium V 51·4	Chromium Cr 52·1	Manganese Mn 55·0	Iron Fe 55·9, Cobalt Co 59, Nickel Ni 59 (Cu)
5		Copper Cu 63·6	Zinc Zn 65·4	Gallium Ga 70·0	Germanium Ge 72·3	Arsenic As 75	Selenium Se 79	Bromine Br 79·95	
6	Krypton Kr 81·8	Rubidium Rb 85·4	Strontium Sr 87·6	Yttrium Y 89·0	Zirconium Zr 90·6	Niobium Nb 94·0	Molybdenum Mo 96·0	—	Ruthenium Ru 101·7, Rhodium Rh 103·0, Palladium Pd 106·5 (Ag)
7		Silver Ag 107·9	Cadmium Cd 112·4	Indium In 114·0	Tin Sn 119·0	Antimony Sb 120·0	Tellurium Te 127	Iodine I 127	
8	Xenon Xe 128	Cæsium Cs 132·9	Barium Ba 137·4	Lanthanum La 139	Cerium Ce 140	—	—	—	— — —
9		—	—	—	—	—	—	—	
10	—	—	—	Ytterbium Yb 173	—	Tantalum Ta 183	Tungsten W 184	—	Osmium Os 191, Iridium Ir 193, Platinum Pt 194·9 (Au)
11		Gold Au 197·2	Mercury Hg 200·0	Thallium Tl 204·1	Lead Pb 206·9	Bismuth Bi 208	—	—	
12	—	—	Radium Rd 224	—	Thorium Th 232	—	Uranium U 239		
HIGHER SALINE OXIDES	R	R_2O	RO	R_2O_3	RO_2	R_2O_5	RO_3	R_2O_7	RO_4
HIGHER GASEOUS HYDROGEN COMPOUNDS					RH_4	RH_3	RH_2	RH	

Much of Mendeleyev's success depended on his placing elements with similar properties in the same group, even though this left occasional gaps in the table. He reasoned, correctly, that the elements that belonged in the gaps had not yet been discovered. The locations of the gaps enabled Mendeleyev to predict with remarkable accuracy the properties of these yet-to-be-found elements.

The table that Mendeleyev developed is in many ways similar to the one we use today. One of the main differences is that Mendeleyev's table lacks the column containing the elements helium through radon. In Mendeleyev's time none of the elements in this column had yet been found because they are relatively rare and because they show no tendency to undergo chemical reactions. Occasionally, Mendeleyev was forced to switch the order of his elements to make the table come out right, placing elements with greater atomic weights ahead of those with smaller weights. The atomic weight of an element is the average weight of all the atoms that form the element. It is determined mostly by the total number of protons and neutrons the atoms contain. Electrons are so much lighter than these nuclear particles that they contribute very little to the weight of an atom. Apparently, listing elements in order of their atomic weights did not always work. It was not until the beginning of the 20th century, with the knowledge gained about the structure of the atom, that the correct way of ordering the elements was discovered and the present periodic table was formulated.

The Nuclear Atom

The key event that led to the modern understanding of the atom was the discovery that atoms are made up of electrons, protons, and neutrons. Thus, despite its name, which derives from the Greek word for "indivisible," the atom could indeed be divided into smaller components.

In April 1897, Joseph John Thompson, professor of physics and director of the Cavendish Laboratory at Cambridge University in England, announced the discovery of the electron. Thompson reported that this tiny particle had a negative electric charge and a mass of about 0.002 that of the lightest atom. Thompson's momentous discovery of a particle of matter smaller than the atom so startled his colleagues that many thought he had been "pulling their legs." It was no joke, however. In Thompson's own words: "The production of electrons essentially involves the splitting up of the atom, [with] a part of the mass of the atom getting free and becoming detached from the original atom—that part being one or more electrons."

Ernest Rutherford, a distinguished New Zealand physicist who had been a pupil of Thompson's and who was a professor of physics at Cambridge University, supplied the next step toward the modern understanding of the atom in 1911. He discovered that the atom had a nucleus and that one of the important particles that occupied the nucleus was the positively charged proton.

As a probe for his study of the atom, Rutherford used the newly discovered phenomenon of radioactivity. Radioactive atoms, like uranium and radium, are unstable, and their nuclei spontaneously disintegrate. One of the products of this disintegration is a massive, positively charged particle called an alpha particle. At the time of its discovery, Rutherford did not know that the alpha particle was the nucleus of a helium atom, consisting of two protons and two neutrons. He therefore used the first letter of the Greek alphabet, *alpha*, to identify this particle and distinguish it from the other products given off by radioactive atoms.

Because the atom was too small to observe directly, Rutherford's brilliant idea was to use alpha particles as projectiles, firing them at atoms and observing how they scattered. This was like firing bullets at a sealed box and deducing the contents of the box by seeing how the bullets bounced. His target atoms were at first gold atoms contained in very thin sheets of gold foil. Gold was used because it is possible to make gold foil that is very thin, often thinner than fine paper. Rutherford observed that although most of the alpha particles passed right through the target, many were deflected at very large angles. Some were even deflected backward, as if they had hit a stone wall. Rutherford was so astonished by this that he compared it to "firing a 15-inch shell at a piece of tissue paper and having it come back and hit the gunner." Because most of the alpha particles went right through the foil, he reasoned that the atom was mainly empty space but that it must contain a small, heavy, positively charged core that was capable of repelling and scattering projectiles fired at it. Rutherford called this massive core the nucleus of the atom.

After Rutherford's discovery of the nucleus, it became obvious that the nucleus of hydrogen, the lightest of the atoms, must play a fundamental role in the structure of all atoms. In 1920, he proposed to call this particle the proton, the name by which it has been known ever since.

Finally, in 1932, the British physicist Sir James Chadwick, who also worked at the Cavendish Laboratory in Cambridge, discovered that yet another particle existed in the nucleus of atoms. This new particle was the neutron. It has a mass close to the mass of the proton, but it has no electric charge.

These fundamental discoveries, coupled with the work of a brilliant young English physicist named Henry Moseley, ultimately led to the reason for Mendeleyev's success with the periodic table. Moseley, just before World War I, had been investigating the X rays given off by various elements. X rays are a very penetrating form of radiation usually produced by accelerating electrons to high speeds and then abruptly stopping them by having them smash into a metal target. The collision causes the target to give off X rays. When different elements are used as targets, the X rays have different properties. Each element has its own set of characteristic X rays. They are almost like a fingerprint of the element. Moseley was able to relate the properties of the X rays to the number of protons contained in the element. He discovered that every element had a different number of protons in the nucleus. The number of protons came to be called the atomic number of the element, represented by the letter Z, and it was always a whole number.

Atoms are normally electrically neutral, with an equal number of electrons and protons. This means, for example, that carbon, with an atomic number of 6, has six protons in its nucleus and six electrons outside the nucleus.

Isotopes

Moseley's experiments demonstrated that what distinguishes one element from another is its atomic number, the number of protons in the nucleus of its atoms, not its atomic weight, which is a measure of the total number of protons and neutrons in the nucleus. The correct way of ordering the elements in the periodic table was, therefore, by their atomic number, and not, as Mendeleyev had thought, by their atomic weight.

Although all the atoms of a given element have the same number of protons, they can have different numbers of neutrons. This explains, for example, why there are three different species of the element hydrogen. Ordinarily, a hydrogen atom has a lone proton in its nucleus and no neutrons. A heavier form of hydrogen, called deuterium, also has a single proton in the nucleus but contains a neutron as well. A still heavier form of hydrogen, known as tritium, has two neutrons in addition to the proton. These three species are called isotopes of the element hydrogen. Yet even though a deuterium atom, because of its extra neutron, weighs twice as much as an ordinary hydrogen atom, its chemical behavior is similar to that of hydrogen, indicating that the number of protons in the nucleus is what determines the behavior of each element.

Like hydrogen, the majority of the elements have isotopes. Some elements have only two isotopes, whereas others can have as many as eight or nine.

The existence of isotopes also explains why the atomic weight is an unreliable indicator of the position of an element in the periodic table. For any element, the atomic weight really measures the average weight of a mixture of its different isotopes. On this basis, it is possible for an element like argon, which has an atomic number of 18, to exist in a mixture of isotopes that have a greater average atomic weight than that of potassium, whose atomic number is 19. The atomic weight in the periodic table is often a number with a decimal fraction. The atomic weight of calcium, Ca, for example, is 40.08. The nucleus of the calcium atom cannot contain 0.08 of a neutron. This number is derived by averaging the weights of the isotopes of calcium that occur together, some of which have more than 20 neutrons in the nucleus.

It is a remarkable fact that the relative percentage, or abundance, of each of the isotopes of any element with more than one isotope can vary depending on where it is found in nature. This means that the atomic weight of an element with several isotopes can also vary, depending on its source. For the first time in history, this variation in atomic weights for 10 important elements—hydrogen, lithium, boron, carbon, nitrogen, oxygen, silicon, sulfur, chlorine, and thallium—has been officially quantified and will be part of all future periodic tables.

The International Union of Pure and Applied Chemistry (IUPAC), an international body of scientists responsible for setting standards in chemistry, published new values for these elements on December 12, 2010, in their journal *Pure and Applied Chemistry.* Based on almost 25 years of work by the IUPAC, the U.S. Geological Survey, and other institutional and university groups, the atomic weights will be expressed as an interval, having a lower and upper boundary of the values found in nature. For example, the atomic weight of nitrogen, instead of its customary value of 14.00674, will be listed as [14.00643; 14.00728], with its lower and upper boundaries separated by a semicolon. This interval expresses the variation in the atomic weight of nitrogen found in the analysis of known substances that contain nitrogen. The new values for the selected elements are shown in side bar.

	New Standard	Traditional Standard
Hydrogen	[1.00784; 1.00811]	1.00794
Lithium	[6.938; 6.997]	6.941
Boron	[10.806; 10.821]	10.811
Carbon	[12.0096; 12.0116]	12.011
Nitrogen	[14.00643; 14.00728]	14.00674
Oxygen	[15.99903; 15.99977]	15.9994
Silicon	[28.084; 28.086]	28.0855
Sulfur	[32.059; 32.076]	32.066
Chlorine	[35.446; 35.457]	35.4527
Thallium	[204.382; 204.385]	204.3833

The differences are small and generally would be important only for scientists doing very precise work. As small as they are, however, modern analytical instruments are quite capable of using these differences to trace the source of certain pollutants or the origin of organic materials containing carbon.

The variation of atomic weights for other elements is also being investigated, and the results should be made available within the next few years.

The Modern Periodic Table

The modern statement of the periodic law is that the chemical and physical properties of the elements vary in a periodic way with their atomic numbers.

The modern periodic table is arranged very much like Mendeleyev's table. The elements are arranged in rows called periods; in each period the elements are arranged in order of their atomic number. The periods are numbered from 1 to 7 from the top row to the bottom row. Below the main body of the table are two long rows of 14 elements each. One of these long groups follows lanthanum (Z = 57) and is known as the lanthanides. The other group follows actinium (Z = 89) and is known as the actinides. These elements actually belong in the main body of the table but are too long to fit conveniently into it.

The vertical columns of the periodic table are called groups. There has been some disagreement about how these should be numbered. In one commonly used system the groups are labeled with Roman numerals and divided into A-groups and B-groups. The IUPAC has officially adopted a system in which the groups are simply numbered in sequence from left to right, using Arabic numerals from 1 to 18. Thus, Group VIIA in the old Roman numeral system is Group 17 in the IUPAC system. The scientific world has still not achieved uniformity in the system used for the periodic table, and most chemists in the United States prefer the more traditional system. We will also use the traditional system in this book, but reference to the periodic table shown on page 6 will quickly translate the heading for a particular column into the IUPAC scheme.

All of the elements within a group have similar chemical properties and are sometimes referred to as families of elements. The elements in the A-groups, or longer groups, are known as representative elements. The elements in the B-groups are called transition elements.

Many of the groups of elements in the periodic table have acquired common names. For example, the elements in group IA, with the exception of hydrogen, are called the alkali metals. The elements in Group IIA are called the alkaline-earth metals, and those in Group VIIA are called the halogens.

What causes this periodic behavior of the elements? Why do the elements within a particular group have similar chemical behavior? The reason is that atoms are attracted to each other by electric forces. The atomic number, the number of positively charged protons in the nucleus, determines how many negatively charged electrons are contained in the atoms of a particular element, and it is the electrons that determine how elements behave and react with one another. The chemical behavior of an element is determined by the way in which the electrons orbiting the nucleus are structured. It was the new quantum physics, developed in the early 20th century by the Danish physicist Niels Bohr, the German physicist Werner Heisenberg, and the Austrian physicist Erwin Schrödinger, that put forward the idea of a complex arrangement of electron orbits or "energy levels" as a way of explaining the bonding properties of elements.

This new quantum physics, based on the idea that matter has properties resembling those of waves, tells us that the electrons in an atom are restricted to certain orbitals. These orbitals, which vaguely resemble the orbits of the planets of our solar system around the sun, are often referred to as shells. The inner shells, closest to the nucleus, are the most stable, and the electrons in these shells are closely held by the attractive force of the nuclear protons. If an electron absorbs energy, it jumps to the next outer orbital. If an electron in an outer orbital gives off energy, it drops to the next inner orbital. Electrons in the outer shells are relatively loosely bound to the nucleus. These electrons may be attracted to other atoms or they may become energetic enough to separate from the atom altogether, leaving behind an atom with a net positive charge that will attract electrons belonging to other atoms.

Strict rules govern how many electrons can occupy any particular shell of an atom. For example, 2 electrons will fill the first shell closest to the nucleus, whereas 8 can occupy the next shell, slightly farther out from the nucleus, and 18 can occupy the shell beyond this. Because each major shell contains various subshells, the exact electron configuration of an atom can become quite complex. The distribution of the electrons in the outer shell of the atom, the one farthest from the nucleus, is the important one, however, because these are the electrons that are exposed to other atoms when the atoms react.

Atoms with similar outer-shell configurations have similar chemical properties. Chemists call the outer shell the valence shell, and the electrons that occupy it are known as the valence electrons. The term *valence* is derived from the Latin word *valent*, which means "strength." The valence electrons determine the chemical "strength" of atoms—their reactivity, or how strongly and in what way they will bind with other atoms. Elements in the same group in the periodic table have the same number of electrons in their outer shells and are therefore said to have the same valence electron configuration. As a result, the chemical and physical properties of the elements in this group will be similar. As the inner shells of an atom become filled with electrons, its outer shell takes on a specific valence configuration that is determined by the rules that govern how many electrons can occupy a particular shell. It is this regularity in the number of electrons that occupy the outer shell that accounts for the periodic behavior shown by the elements as the atomic number increases. Other properties, such as the size of an atom, are also determined by the number of shells it contains. For example, the radius of the atoms of the elements in a particular group in the periodic table tends to increase from the top of the group to the bottom.

Elements whose shells are completely full are extremely stable and seem to react with almost nothing else. The elements of

Group VIIIA, for example, the so-called noble gases, all have complete shells and are the most chemically inert elements that exist. A complete shell of electrons is so energetically stable that atoms with incomplete shells will tend to react with other atoms in a manner that will complete these shells. In other words, atoms react in order to attain a "noble gas" configuration. In moving from left to right across a horizontal row, or period, within the periodic table, there is a transition from elements that are metals to those that are nonmetals. Metals, which generally have few electrons in their outermost valence shells, tend to lose electrons when they react, so that they reach a state in which they have fewer shells, all of which are completely filled with electrons. Nonmetals, whose valence shells are almost completely filled, tend to accept electrons to fill these shells and stabilize their configuration. In both cases, the tendency is to assume a completely filled valence shell, approximating that of a noble gas.

The periodic table, then, is a map of the way in which electrons arrange themselves in the atoms of a particular element. As you go down a column within a group, all the elements of that group have the same number of valence electrons. As you go across a row, from left to right, electrons are being added to a shell. The ability to predict the chemical behavior of an element, based on the row and column in which it is found, makes the periodic table an indispensable reference tool for scientists. Open a chemistry textbook and the chances are that there will be a periodic table, often in bright colors, printed on the inside cover of the book. Its constant use by chemists emphasizes the central role the periodic table plays in making sense of what otherwise might be a chaotic jumble of facts about the elements and their many molecular combinations.

The chemical group of each of the elements described in the sections that follow is listed directly below the chemical symbol of the element. The similarity of chemical properties of elements in the same group should be as apparent to you as it was to Mendeleyev or to any chemist who uses the periodic table for information and research.

HOW ELECTRONS OCCUPY ATOMIC SHELLS

	Scientific name	Permitted subshells	Maximum electrons in subshell	Maximum electrons in shell
Shell closest to nucleus	K	1s	2	2
Next shell farther out	L	2s	2	8
		2p	6	
Next outer shell	M	3s	2	18
		3p	6	
		3d	10	
Next outer shell	N	4s	2	32
		4p	6	
		4d	10	
		4f	14	

Hydrogen

Atomic Number 1

Chemical Symbol H

Group IA

IA																	VIIIA
H	IA											IIIA	IVA	VA	VIA	VIIA	He
Li	Be											B	C	N	O	F	Ne
Na	Mg	IIIB	IVB	VB	VIB	VIIB	VIIIB			IB	IIB	Al	Si	P	S	Cl	Ar
K	Ca	Sc	Ti	V	Cr	Mn	Fe	Co	Ni	Cu	Zn	Ga	Ge	As	Se	Br	Kr
Rb	Sr	Y	Zr	Nb	Mo	Tc	Ru	Rh	Pd	Ag	Cd	In	Sn	Sb	Te	I	Xe
Cs	Ba	*La	Hf	Ta	W	Re	Os	Ir	Pt	Au	Hg	Tl	Pb	Bi	Po	At	Rn
Fr	Ra	†Ac	Rf	Db	Sg	Bh	Hs	Mt	Ds	Rg	Cn	Uut	Uuq	Uup	Uuh	Uus	Uuo

*	Ce	Pr	Nd	Pm	Sm	Eu	Gd	Tb	Dy	Ho	Er	Tm	Yb	Lu
†	Th	Pa	U	Np	Pu	Am	Cm	Bk	Cf	Es	Fm	Md	No	Lr

Hydrogen is the simplest of all the atoms. It consists of nothing more than a single proton, which serves as its nucleus, circled by a single electron. Its simplicity helps to explain why it is by far the most abundant element in the universe. Huge quantities exist in interstellar space, and it is the predominant element in stars. It has such a dominant position among the other elements that it makes up an astonishing 93 percent of all the atoms in the universe.

It is a very ancient element, having been formed shortly after the Big Bang, the moment when the universe is thought to have exploded into existence. All the other elements were made either by nuclear reactions taking place in the core of burning stars or by the catastrophic explosions called supernovas that are sometimes produced when stars die.

Given the major role played by hydrogen in the universe, it is surprising to learn that there is very little hydrogen gas in the Earth's atmosphere. If you were to take 100 million liters of air, only about 5 liters would be hydrogen. Hydrogen is a very light gas. The gas weighs so little that the gravitational pull of the Earth was not strong enough to have prevented most of the hydrogen that was once present in the air from escaping into outer space. Some of the larger planets, such as Saturn and Jupiter, exert a much greater gravitational attraction and contain considerably more hydrogen gas in their atmospheres.

Much of the hydrogen still found on Earth is bound up in the water molecules that form our great oceans and seas. About 3 percent of the Earth's crust is made up of hydrogen atoms.

Under normal conditions, hydrogen gas is a diatomic molecule. This means that a hydrogen molecule is made up of two atoms of hydrogen. The chemical symbol of the gas is written as H_2 to show the coupling of the atoms. The gas has no odor or taste and is completely colorless. It is, however, an extremely flammable gas. When hydrogen burns in air, it combines very vigorously with oxygen present in the air to form large amounts of heat. The combination of hydrogen and oxygen also produces water. It is this reaction that gave birth to the name of hydrogen, which is derived from the Greek words *hydro*, or "water," and *genes*, or "creator."

The discovery of hydrogen is usually credited to the English chemist Henry Cavendish (1731–1810), after whom Cambridge University's Cavendish Laboratory is named. Cavendish was about as eccentric a scientist as there ever was. The son of English aristocrats, he spent almost none of his huge inheritance and devoted his entire life to science. He worked out a value for G, the gravitational constant, and accurately calculated the mass of the Earth, which confirmed Isaac Newton's theory of universal gravitation. But whenever Cavendish had to talk to anyone, he was shy to the point of stammering, and he never spoke to women. He gave instructions to his female servants by handwritten notes, and he would fire them instantly if they did not stay hidden from his sight when he walked around his house.

Thanks to a large inheritance from his family, the eccentric English chemist Henry Cavendish was able to devote his entire life to science, experimenting in a wide range of areas. He is best known for his discovery of hydrogen in 1766.

In 1766, Cavendish produced hydrogen gas by adding some zinc metal to an acid. He recognized that the "inflammable air," as it was then called, being given off by the reaction was a distinct substance, and he identified it as an element. This method is still used to produce small quantities of hydrogen gas in laboratories. A few pieces of zinc, or some iron filings, added to a test tube containing some hydrochloric acid will produce bubbles of hydrogen gas. When large amounts of hydrogen are needed, most laboratories and manufacturing plants use cylinders of the compressed gas.

Hydrogen is well known for being a gas that is lighter than air. A balloon filled with hydrogen will immediately start rising when released and float away. Its low density, the smallest of any gas, gives it this great lifting power in air. Hydrogen was once used to keep blimps and manned balloons afloat. After a spark set the German blimp *Hindenburg* aflame in 1937, however, the use of hydrogen for filling airships was abandoned. It has since been replaced by helium, which is slightly denser but far safer because it is nonflammable.

The most common compound of hydrogen is water. In the creation of this compound, two atoms of hydrogen combine with one atom of oxygen to form the familiar water molecule H_2O. But hydrogen is also contained in an almost uncountable number of organic, or carbon-containing, compounds and biological compounds present in living organisms. It is often combined directly with carbon in organic molecules. Among the immense variety of organic molecules in which hydrogen is linked chemically to carbon are hydrocarbons, the long, chainlike molecules found in natural gas and oil that, when broken apart, release the energy we use to run our power plants and automobiles. Another group of organic compounds is the carbohydrates, which consist of molecules of hydrogen, carbon, and oxygen found in sugars and starchy foods that supply humans and plant-eating animals with energy. Hydrogen compounds are also found in perfumes, dyes, pesticides, DNA, and proteins. The list goes on and on.

Hydrogen is usually prepared commercially by decomposing water. In a famous reaction called the "water gas reaction," steam is passed over hot carbon in the form of coke, although methane gas is sometimes substituted for the coke. Methane, whose chemical formula is CH_4, contains four hydrogen atoms per molecule. And if the water used is in the form of superheated steam, the hydrogens from both the methane and the water are freed from

The space shuttle Endeavour *blasts off in 1992. Large quantities of liquid hydrogen are used to power the shuttle.*

their molecules to form hydrogen gas. The steam that reacts with coke produces gaseous carbon monoxide and hydrogen. These two gases can be physically separated, but often the combination of the gases, called "water gas," is used industrially as a fuel.

Commercially produced hydrogen gas has many uses in the chemical and food industries. Probably its most important use in the chemical industry is for the manufacture of ammonia, an ingredient of fertilizers. For the preparation of food, large quantities of hydrogen are used in a process called hydrogenation. In this process, hydrogen is added to a liquid vegetable oil, converting it to a solid such as margarine. Because it contains much less cholesterol, a fatty substance that tends to clog blood vessels, margarine is used as a substitute for butter, which is an animal fat. The aerospace industry also requires large quantities of liquid hydrogen for the manufacture of rocket fuels. When mixed with oxygen in a propulsion chamber, the hydrogen burns to produce hot steam that supplies the thrust for rocket propulsion.

Hydrogen has two isotopes of interest. The isotope containing a single neutron in addition to the single proton that normally constitutes the nucleus of hydrogen is known as hydrogen-2, or deuterium (D); the isotope with two extra neutrons is hydrogen-3, or tritium (T). These isotopes have the distinction of being the only isotopes of any element that have individual names.

Deuterium is a stable isotope that is heavier than ordinary hydrogen because of the extra neutron in its nucleus. It is often known as "heavy" hydrogen. Like its lighter cousin, deuterium combines with oxygen to form water, which in this case is known as "heavy" water, or D_2O. The natural abundance of deuterium is only about 0.02 of 1 percent, which means that about 1 of every 6,000 water molecules found in the oceans and lakes of the Earth is heavy water. Deuterium is usually separated from water by a process known as electrolysis. Here the water is decomposed into hydrogen and oxygen gas by passing an electric current through it. A small amount of the hydrogen will be deuterium, and the two gases are then separated using the difference in their atomic weights.

Deuterium is unusual in that its chemistry differs somewhat from that of hydrogen. For example, water that contains more than 40 percent D_2O is toxic. This difference in the chemical behavior of the isotopes of hydrogen from the conventional element is caused by the different weights of the atoms and is almost unique among the elements.

Heavy water is chiefly used as a moderator in nuclear reactors. A moderator serves to "moderate," or slow down, the

A hydrogen bomb is based on the same process—nuclear fusion—by which the sun produces the light and heat that sustain life on Earth.

neutrons released in the reactor during the process of nuclear fission. Surprisingly, uranium atoms will undergo fission more easily if hit by slow-moving neutrons, and heavy water greatly increases the efficiency of the nuclear reactions in which uranium produces energy in a reactor.

Tritium, which contains one more neutron than deuterium, is an even heavier isotope of hydrogen. Its extra neutron is apparently enough to make tritium unstable, so that it is radioactive, with a relatively short half-life of 12.26 years. The half-life of a radioactive substance is the time required for half of its atoms to decay into atoms of lighter elements. Despite its short half-life, tritium is still found on the Earth because it is constantly being produced. High-energy particles, called cosmic rays, are constantly coming in from outer space and bombarding the upper layers of the Earth's atmosphere. This bombardment produces nuclear reactions that create tritium.

Like ordinary hydrogen, tritium reacts with the oxygen in the atmosphere to form T_20, a radioactive "water" molecule. This radioactive water constantly enters the Earth's seas and lakes in the form of slightly radioactive rain. Fortunately, its half-life is short enough to prevent a hazardous buildup of radioactivity.

Deuterium and tritium have become key fuels in the attempt to create a device that uses nuclear fusion to produce energy. A fusion reaction is a process in which the nuclei of two atoms are brought into sufficiently close proximity for them to fuse together, almost like two water drops, forming a new, larger nucleus and liberating large amounts of energy in the process. The fusion of hydrogen nuclei is the process by which the sun produces the light and heat energy that sustains life on Earth. The hydrogen bomb, which also uses the fusion of hydrogen isotopes, is based on the huge amounts of energy available from such a reaction.

A bomb, however, is a reaction out of control. Scientists today are trying to produce a controlled fusion reaction, and one that will sustain itself. The hope is to create a fusion machine that produces more energy than is required to make the reaction work. The energy yield from such a device, if it could be controlled, would be so great that major research efforts are under way all over the world to overcome the enormous scientific and engineering difficulties facing workers in this field. One of the major problems is that the nuclei of the hydrogen atoms that scientists want to fuse have to be heated to temperatures of hundreds of millions of degrees. These high temperatures are needed to give the hydrogen nuclei enough energy to overcome their natural electrical repulsion from one another, because both nuclei are positively charged, and make them fuse together. The reason that deuterium and tritium are the atoms of choice in creating such a fusion reactor is that the temperature required to overcome their repulsion is lower than that needed for any other feasible fuel. Although the concentration of deuterium in the Earth's oceans is low, the Earth has so much water that the deuterium available for use in a fusion reactor would be almost unlimited.

A bomb is a reaction out of control. Scientists today are trying to produce a controlled fusion reaction and one that will sustain itself.

Helium

IA	IA	IIIB	IVB	VB	VIB	VIIB	VIIIB			IB	IIB	IIIA	IVA	VA	VIA	VIIA	VIIIA
H																	He
Li	Be											B	C	N	O	F	Ne
Na	Mg											Al	Si	P	S	Cl	Ar
K	Ca	Sc	Ti	V	Cr	Mn	Fe	Co	Ni	Cu	Zn	Ga	Ge	As	Se	Br	Kr
Rb	Sr	Y	Zr	Nb	Mo	Tc	Ru	Rh	Pd	Ag	Cd	In	Sn	Sb	Te	I	Xe
Cs	Ba	*La	Hf	Ta	W	Re	Os	Ir	Pt	Au	Hg	Tl	Pb	Bi	Po	At	Rn
Fr	Ra	†Ac	Rf	Db	Sg	Bh	Hs	Mt	Ds	Rg	Cn	Uut	Uuq	Uup	Uuh	Uus	Uuo

*	Ce	Pr	Nd	Pm	Sm	Eu	Gd	Tb	Dy	Ho	Er	Tm	Yb	Lu
†	Th	Pa	U	Np	Pu	Am	Cm	Bk	Cf	Es	Fm	Md	No	Lr

Atomic Number 2

Chemical Symbol He

Group **VIIIA—The Noble Gases**

He

Add a second proton and two neutrons to the hydrogen nucleus and you have helium. In the periodic table, helium heads the column known as the noble gases, so called because they are chemically inert and give the impression of keeping aloof from other elements. Like all the gases in this group, helium is colorless and odorless. It is so inactive that it will not even combine with itself. In fact, the basic unit of helium gas consists of a single atom. Helium's inertness makes it one of those rare elements that always exists in its pure elemental form. No stable compound of helium has ever been made.

Helium is second only to hydrogen as the most abundant element in the universe. It makes up about 7 percent of all the existing atoms. Together, hydrogen and helium form an astonishing 99.9 percent of elements in the universe. Helium, however, is relatively scarce on Earth and ranks a poor sixth among the gases that make up the air.

Interestingly, helium was discovered on the sun before being identified on Earth. In 1868 the French astronomer Pierre Janssen (1824–1907) happened to notice an unusual and unexpected spectral line during his analysis of the spectral emission of the sun. Spectral emission lines refer to the wavelengths of light given off by each element when it is heated to a fairly high temperature. Every element gives off colors that are unique to it. Scientists use these colors to identify an element in the same way that a police officer uses fingerprints to identify a suspect. The yellow light given off by the sodium vapor in lamps used for highway illumination is an example of the pure light that an element produces.

Janssen correctly reasoned that the strange spectral line he detected in the sun's light must be caused by an element that had never before been identified. He named this new element helium, from the Greek *helios*, which means "sun." In 1895, the Scottish chemist William Ramsay discovered helium gas in a sample of uranium ore on Earth.

The helium in the sun is produced by the fusion of hydrogen. Scientists often refer to this process as "hydrogen burning." It is this reaction that supplies the energy that the sun radiates out into space. At the high temperatures and pressures found in the interior of the sun, a number of nuclear reactions are present that have the net effect of fusing protons into a helium nucleus. Temperatures as high as 10 million degrees Celsius and pressures 100 times that of the atmospheric pressure on Earth are required to start these reactions. Luckily, even at these extreme conditions, hydrogen burns rather slowly. The sun has been burning hydrogen for some 5 billion years and is expected to continue doing so for the next 5 billion years.

On Earth, almost all the helium is found in natural gas and results from the decay of such radioactive isotopes as uranium and radon. These radioisotopes spontaneously emit helium nuclei as they decay. The more common name for these helium nuclei is alpha particles, so called by their discoverer, Sir Ernest Rutherford, a professor of physics at the Cavendish Laboratory at Cambridge University in England. Rutherford did not at first realize that alpha particles were helium nuclei and therefore assigned the Greek letter "alpha" to them, in much the same way that unknowns are generally called x in algebra.

Some 3 billion cubic feet of helium is made commercially each year in the United States alone. Although natural gas, an important source of energy, consists primarily of methane, it also contains helium in concentrations as high as 0.3 percent. The helium is separated from the methane and other contaminants by a process known as fractional distillation. Fractional distillation is a technique of separating a mixture of liquids using the differences in their boiling points. Helium is a difficult gas to liquefy, since its boiling point of –268.9°C is lower than that of any other gas. If natural gas is cooled, all of the other gases in it will liquefy first, leaving only helium as a gas.

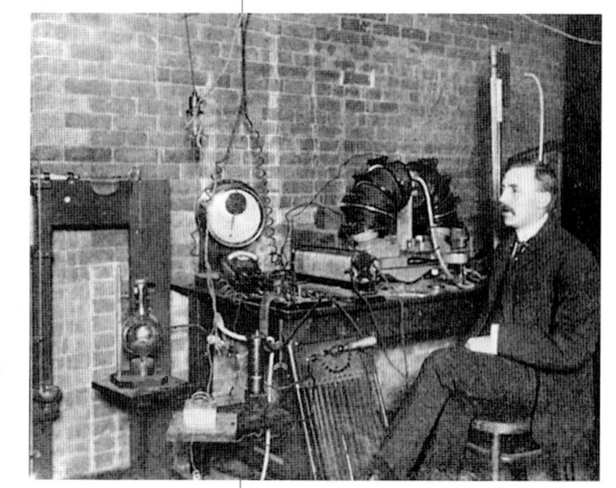

Sir Ernest Rutherford in his laboratory at Cambridge University. Rutherford did not realize at first that the mysterious particles released during the decay of such radioactive isotopes as uranium and radon were actually helium nuclei.

Helium is a light gas with a low density and is therefore useful for its buoyancy in air. Although it is denser than hydrogen, and therefore has less lifting power than hydrogen, it has the major advantage of not being flammable or toxic. Blimps and toy balloons are commonly filled with helium to make them

Because helium is lighter than air, it can be used to make blimps and balloons float. Although hydrogen has more lifting power than helium, its flammability makes it unsafe to use.

float in air. Divers working at high pressures underwater usually breathe "air" that is a mixture of oxygen and helium. Helium is substituted for nitrogen in this mixture because it is less soluble than nitrogen and therefore less likely to dissolve in the bloodstream. This offers better protection against one of the major hazards of diving, called the "bends." If a diver returns to the surface too quickly after a dive, the relatively lower pressure at the surface than deep underwater causes dissolved gases to bubble out of solution in the blood. The effect is similar to the frothing in a bottle of soda water when the cap is removed. In the body, the bubbles of the gases released in this way often get trapped in the joints of the body, causing extreme pain and often making it impossible for the diver to straighten up, which is why this condition is aptly named the bends.

An amusing side effect of breathing helium is a change in the pitch and quality of one's voice. Because helium is less dense than air, the vocal cords produce sounds of higher pitch than with air and make the speaker sound like Donald Duck.

A mixture of helium and neon gas produced the first gas laser ever made. The word *laser* is an acronym for *l*ight *a*mplification by *s*timulated *e*mission of *r*adiation. Lasers produce intense and very narrow beams of light that are sharply focused on a single color. One of the major applications of the helium–neon laser is in supermarkets, where it is used to read the bar codes of food labels during checkout.

Since World War I the United States has maintained a stockpile of helium. We now store some 32 billion cubic feet of helium gas in natural caverns. Although helium is no longer critical for military purposes, it still has many vital scientific applications. It is used for weather balloons that can also study the upper layers of the atmosphere. Many of the world's large particle accelerators, built to investigate the structure of matter, use liquid helium to cool their superconducting magnets. Astronomers also use liquid helium to cool many of their detectors. This removes the interference called thermal "noise" and makes it easier and more reliable to receive data from distant galaxies.

Lithium

IA	IA	IIIB	IVB	VB	VIB	VIIB	VIIIB			IB	IIB	IIIA	IVA	VA	VIA	VIIA	VIIIA
H																	He
Li	Be											B	C	N	O	F	Ne
Na	Mg											Al	Si	P	S	Cl	Ar
K	Ca	Sc	Ti	V	Cr	Mn	Fe	Co	Ni	Cu	Zn	Ga	Ge	As	Se	Br	Kr
Rb	Sr	Y	Zr	Nb	Mo	Tc	Ru	Rh	Pd	Ag	Cd	In	Sn	Sb	Te	I	Xe
Cs	Ba	*La	Hf	Ta	W	Re	Os	Ir	Pt	Au	Hg	Tl	Pb	Bi	Po	At	Rn
Fr	Ra	†Ac	Rf	Db	Sg	Bh	Hs	Mt	Ds	Rg	Cn	Uut	Uuq	Uup	Uuh	Uus	Uuo

*	Ce	Pr	Nd	Pm	Sm	Eu	Gd	Tb	Dy	Ho	Er	Tm	Yb	Lu
†	Th	Pa	U	Np	Pu	Am	Cm	Bk	Cf	Es	Fm	Md	No	Lr

Li

Lithium is the element that follows helium in the periodic table. The difference in the chemical and physical properties of these two elements could not be more striking. Lithium is a metal, an extremely reactive one. It reacts vigorously with water, for example, to produce hydrogen gas and with oxygen to form lithium oxide (Li_2O). To prevent these reactions from occurring spontaneously in air, lithium is usually isolated from its environment by storing it immersed in oil or kerosene.

Lithium is the lightest metal of the group known in the periodic table as the alkali metals. Lithium is at the head of this group. All these metals are very reactive. Lithium metal is so soft that it

Lithium is such a soft metal that it can be cut with a sharp knife.

Atomic Number **3**

Chemical Symbol **Li**

Group **IA—The Alkali Metals**

Lithium was discovered in 1817 by a Swedish chemist during a routine chemical investigation of some minerals from a mine in Sweden.

can be cut with a sharp knife. It is never found in its "free," or pure, state in nature, since it reacts so readily with air and water. When purified, it has a beautiful silvery-white color.

It is startling to think of a metal floating on water, but the density of lithium is so low that it actually does float. When carefully placed on the surface of water, a small piece of lithium will whirl about erratically, very much like a small boat out of control.

Lithium was discovered in 1817 by the Swedish chemist Johan August Arfvedson during a routine chemical investigation of some minerals from a mine in Sweden. It was named for the Greek word *lithos*, which means "stone."

All the alkali metals are reactive, but their reactivity increases with their atomic number. This means that these metals become more reactive as you go down their column in the periodic table. This can be explained by noticing that the "valence" electron, the electron occupying the outermost shell of the atom, is farther from the nucleus in large atoms. The farther away the electron is from the nucleus, the less tightly it is bound to the atom and the more likely it is to be removed from the atom in a chemical reaction. Thus, for example, when rubidium and cesium—the heaviest of the alkali metals and the two members of this group with the largest atoms—react with water, the reaction proceeds with explosive violence, rather than with a simple bubbling of hydrogen.

The most common commercial source of lithium is the mineral spodumene. A compound called lithium carbonate (Li_2CO_3), obtained from this mineral, is the starting point for obtaining metallic lithium and most of its important compounds. The pure metal itself, for example, is obtained by the electrolysis of lithium chloride (LiCl), which is prepared from lithium carbonate. Electrolysis is a process in which electricity is used to produce a chemical change in a substance. In the electrolysis of lithium chloride, two electrodes, one called an anode and the other a cathode, are placed in a vat of molten lithium chloride. The electrodes are connected to a source of electricity and large electric currents are passed through the molten compound. Lithium metal forms at the cathode and is removed for further processing.

The use of lithium metal has assumed some commercial importance in recent years. It is, for example, combined with aluminum to form a low-density, structurally strong alloy for use in aircraft and spaceships. Lithium metal is also used as the positive terminal, or anode, in the small batteries used in cameras, pacemakers, and calculators. In addition to being lighter than the standard dry cell, these batteries produce a higher voltage (3 volts versus 1.5 volts).

Lithium hydroxide (LiOH) is an important compound of lithium obtained from the carbonate. It is a strong base, which means that when dissolved in water, it produces a high concentration of hydroxide $(OH)^-$ ions. An ion is an atom or molecule that has a net electric charge because it contains fewer or more electrons than protons. If electrons are stripped away from an atom, a positively charged ion is formed. If electrons are gained, a negatively charged ion is formed. Ions are highly reactive because their electric charge attracts other atoms, with which they combine in an effort to lose their electron imbalance and once again become electrically stable. When lithium hydroxide is heated with a fat, it produces a soap called lithium soap. (All soaps are essentially prepared by combining animal fats with a strong base.) Lithium soap has the ability to thicken oils and so is used commercially to manufacture lubricating greases.

Lithium hydroxide is also a very efficient and lightweight purifier of air. In confined areas—aboard spacecraft or submarines, for example—the concentration of carbon dioxide from exhaled air can often approach unhealthy or even toxic levels. Lithium hydroxide absorbs the carbon dioxide from the air by reacting with it to form lithium carbonate.

In its purified form, lithium carbonate has recently been used to treat patients with the severe mental illness known as bipolar disorder. More than 3 million prescriptions for this compound are filled yearly to help patients with this disorder. The role of lithium carbonate in the complex chemical reactions that determine brain dysfunction is not known, but it is thought to affect the way in which brain cells respond to certain hormones and to the complex biological molecules known as neurotransmitters, which assist the transmission of messages along nerve networks and can greatly affect moods and behavior.

One of the isotopes of lithium, lithium-6 (so designated because its nucleus contains three neutrons in addition to three protons), played a major role in the production of the hydrogen bomb. The hydrogen bomb makes use of the thermonuclear fusion of deuterium and tritium to produce vast amounts of energy. In order to fuse the nuclei of these isotopes of hydrogen, however, temperatures as high as millions of degrees are required. The problem in creating the hydrogen bomb was to fabricate a device that could produce these high temperatures and supply enough deuterium and tritium to make an effective bomb.

During World War II, the Hungarian-born American physicist Edward Teller saw an ingenious way to construct such a bomb using a compound of lithium-6 and deuterium that is

The American physicist Edward Teller used a compound of lithium—lithium deuteride—to help construct a bomb far more powerful than the original atomic bombs exploded over Hiroshima and Nagasaki.

known as lithium deuteride. The deuteride is a solid and therefore contains a large amount of deuterium in a small volume. The important feature of this compound, however, is that when lithium-6 is subjected to bombardment by neutrons, it undergoes a nuclear reaction to form tritium. Teller reasoned that if an ordinary uranium fission bomb—the original atomic bomb—was surrounded with a layer of lithium deuteride, the explosion of the uranium bomb would not only supply the neutrons needed to transform lithium-6 into tritium but also produce the temperatures needed to fuse the tritium with deuterium. The idea worked brilliantly, and, for good or evil, superbombs with megaton capabilities were created.

Lithium has the highest heat capacity of any element, more than twice that of water, which means that it can absorb large amounts of heat with only a slight increase in its own temperature.

Lithium plays another role in the world of nuclear energy. It has the highest heat capacity of any element, more than twice that of water, which means that it can absorb large amounts of heat with only a slight increase in its own temperature. This property makes lithium an ideal heat-transfer material, and it is being used in many experimental nuclear reactors to absorb the heat produced by the fissioning of uranium.

IA	IA	IIIB	IVB	VB	VIB	VIIB	VIIIB			IB	IIB	IIIA	IVA	VA	VIA	VIIA	VIIIA
H																	He
Li	**Be**											B	C	N	O	F	Ne
Na	Mg											Al	Si	P	S	Cl	Ar
K	Ca	Sc	Ti	V	Cr	Mn	Fe	Co	Ni	Cu	Zn	Ga	Ge	As	Se	Br	Kr
Rb	Sr	Y	Zr	Nb	Mo	Tc	Ru	Rh	Pd	Ag	Cd	In	Sn	Sb	Te	I	Xe
Cs	Ba	*La	Hf	Ta	W	Re	Os	Ir	Pt	Au	Hg	Tl	Pb	Bi	Po	At	Rn
Fr	Ra	†Ac	Rf	Db	Sg	Bh	Hs	Mt	Ds	Rg	Cn	Uut	Uuq	Uup	Uuh	Uus	Uuo

*	Ce	Pr	Nd	Pm	Sm	Eu	Gd	Tb	Dy	Ho	Er	Tm	Yb	Lu
†	Th	Pa	U	Np	Pu	Am	Cm	Bk	Cf	Es	Fm	Md	No	Lr

Beryllium

Atomic Number **4**

Chemical Symbol **Be**

Group IIA—The Alkaline-Earth Metals

Be

Following lithium in the periodic table is the element beryllium. Beryllium heads the group of elements known as the alkaline-earth metals. The most abundant elements in this group are calcium and magnesium, but they all share common chemical and physical properties. Although several forms of beryllium are found in the Earth's crust, it is a relatively scarce element, ranking 32nd in order of the relative abundance of the elements.

Beryllium in its pure state is a fairly hard, gray–white metal. It is one of the lightest of the metals, but hard enough to scratch glass. Like all the metals that make up the alkaline-earth group, it is much too chemically reactive to be found in its free state. It readily reacts, for example, with oxygen and many other elements. Its principal natural source is the mineral beryl, from which beryllium gets its name. Beryllium was formerly known as "glucinum" because beryllium, like many of its compounds, has a sweet taste. This name is now obsolete.

Crystals of this mineral are quite beautiful and often very valuable. Both emerald (green) and aquamarine (blue) are naturally

The brilliant green color of this emerald comes from the presence of beryl, the mineral that is the principal source of beryllium.

In 1932, the English physicist James Chadwick discovered neutrons by bombarding beryllium with alpha particles.

occurring precious forms of beryl. When cut and polished, these crystals supply the gems that are used in many expensive bracelets and necklaces.

The mineral known as beryl is actually a complex compound of beryllium, silicon, and oxygen. Its scientific name is beryllium aluminum silicate. It is usually found, like most of the alkaline-earth metals, in various mineral deposits distributed over Brazil, Argentina, and the United States. Crystals of beryl several feet long and weighing as much as a thousand pounds have been found. More commonly, however, these crystals are quite small.

Beryllium was discovered in 1798 by the French chemist Louis-Nicolas Vauquelin, who was investigating the structures of beryl and emerald. The two gemstones were long thought to be different minerals.

Metallic beryllium in commercial quantities is made from the electrolysis of molten beryllium chloride ($BeCl_2$). The pure metal was not commercially available until 1957. Being transparent to X rays, it is used as a sturdy window for X-ray tubes.

Beryllium is often combined with other metals to form special alloys. One of these is the beryllium–copper alloy known as beryllium bronze, a fairly hard metal with the unusual property of not giving off sparks when struck. This makes it a valuable material for the electrical contacts and hammers employed in explosive environments. These might be chemical laboratories using hydrogen or factories manufacturing rocket fuel.

Great care must be used in working with beryllium compounds because they are quite toxic. Exposure and inhalation of finely powdered beryllium compounds, such as beryllium oxide, can lead to a painful and fatal disease known as berylliosis.

When beryllium is bombarded with alpha particles, it undergoes a nuclear reaction that causes it to emit neutrons. It was this reaction that led to the discovery of the neutron by the English physicist James Chadwick in 1932. This reaction is still used today as a convenient source of neutrons.

Beryllium mirrors will play a major role in the James Webb Space Telescope, the next-generation successor to the Hubble Space Telescope. Unlike the Hubble, which detected visible light, the Webb will work in the infrared region and will be able to "see," unimpeded by clouds of dust, regions further back in space and time than is now possible. Webb's huge primary mirror, which will be seven times larger than the Hubble's, is made of 18 hexagonal segments of beryllium, ideally suited to withstand the extreme temperatures of outer space.

IA	IA	IIIB	IVB	VB	VIB	VIIB	VIIIB			IB	IIB	IIIA	IVA	VA	VIA	VIIA	VIIIA
H																	He
Li	Be											B	C	N	O	F	Ne
Na	Mg											Al	Si	P	S	Cl	Ar
K	Ca	Sc	Ti	V	Cr	Mn	Fe	Co	Ni	Cu	Zn	Ga	Ge	As	Se	Br	Kr
Rb	Sr	Y	Zr	Nb	Mo	Tc	Ru	Rh	Pd	Ag	Cd	In	Sn	Sb	Te	I	Xe
Cs	Ba	*La	Hf	Ta	W	Re	Os	Ir	Pt	Au	Hg	Tl	Pb	Bi	Po	At	Rn
Fr	Ra	†Ac	Rf	Db	Sg	Bh	Hs	Mt	Ds	Rg	Cn	Uut	Uuq	Uup	Uuh	Uus	Uuo

*	Ce	Pr	Nd	Pm	Sm	Eu	Gd	Tb	Dy	Ho	Er	Tm	Yb	Lu
†	Th	Pa	U	Np	Pu	Am	Cm	Bk	Cf	Es	Fm	Md	No	Lr

Boron is a hard, brittle, nonmetallic element that stands at the head of the elements that make up Group IIIA in the periodic table. All the other elements of this group, which include aluminum, are metals. Boron is a rather scarce element, making up only about 0.0003 percent of the Earth's crust by mass.

In its pure state, it is an extremely hard black crystal. It was first identified as part of a compound in 1808 by the English chemist Sir Humphrey Davy and two French chemists, Joseph-Louis Gay-Lussac and Louis-Jacques Thenard. A pure sample of the element was not isolated until 1909, by the American chemist William Weintraub.

Boron exists in many allotropic forms in nature, but in 2009 a team working under Artem R. Oganov at Stony Brook University in New York produced a form of the element that had not been seen before. Working at high temperatures and pressures, they created an ionic crystal of boron, a structure that is usually produced by the attraction between two different kinds of atoms that are electrically charged. In the case of boron, two different clusters of boron atoms, with each cluster behaving like a charged ion,

The English chemist Sir Humphry Davy identified boron in 1808. He was also responsible for isolating seven other elements.

Boric acid—a compound of boron—is used to make a special heat-resistant type of glass, borosilicate glass, which is used to manufacture glass baking dishes.

generated the electric force holding the crystal together. The new form is almost as hard as diamond.

Although boron is much less reactive than the elements that immediately precede it in the periodic table, it is never found in its pure form in nature. Instead, it is usually combined with oxygen, water, and sodium in a compound called borax. Borax, the most important compound of boron and the substance that gives boron its name, is found in dry lake beds in the southwestern part of the United States. Borax is used mainly as a cleaning agent and water softener. Water is said to be "hard" if it contains some alkaline-earth ions such as magnesium and calcium, which combine with soap to produce a scumlike precipitate that can cause the familiar rings around sinks and tubs. When water is softened, the magnesium and calcium are removed and replaced with relatively harmless sodium and potassium.

Another common and important boron compound is boric acid. It is made by heating borax with either hydrochloric or sulfuric acid. Boric acid is a rather weak acid that has some antiseptic properties that make it useful as an eyewash. Industrially, it is used to make a special heat-resistant type of glass, known as borosilicate glass, which usually carries the commercial name of Pyrex. The most common use of Pyrex is in the kitchen, where Pyrex glass baking dishes and measuring cups are used because they can withstand rapid changes in temperature without cracking.

Boron plays a very important role in the design and utilization of nuclear reactors. It became apparent quite early in the search for ways to produce energy from the nucleus that boron is a very efficient absorber of neutrons. The neutron, one of the fundamental particles that make up the nucleus of an atom, plays a crucial role in a nuclear reactor. When uranium-235, the most common fuel used in reactors, absorbs a neutron, it splits or "fissions" into two fragments, releasing energy and more neutrons. These new neutrons can then fission more uranium atoms and start a chain reaction. To prevent this chain reaction from running wild and producing an explosion, the number of neutrons available for fissioning must be controlled. Nuclear engineers commonly use boron "rods" that can be lowered into a reactor to absorb the neutrons and so control the power being produced in the reactor. These rods are called control rods.

Boron has also become important in the manufacture of transistors. It is hard to conceive of life today without calculators, computers, and VCRs. All of these instruments use transistors, which are silicon or germanium chips to which impurities, such as boron, have been added in a carefully controlled manner.

When silicon atoms combine with each other to form a solid crystal of silicon, the outer electrons of all of these atoms share in forming the chemical bonds that hold the crystal together. Electrons are found in concentric "shells" about the nucleus of any atom, and there are certain rules, called quantum rules, that determine the number of electrons in each shell. The electrons found in the very outermost, or valence, shell usually determine how an atom reacts chemically, and it is these electrons that are used to form chemical "bonds" between atoms. These valence electrons hold the silicon crystal together.

When a small quantity of an element such as boron is added to a crystal of silicon, boron atoms take over some of the sites that were formerly occupied by silicon atoms in the crystal. Chemists usually refer to the addition of an impurity of this kind by saying that the silicon has been "doped" with boron. However, boron has only three outer electrons, whereas silicon has four. Consequently, the site occupied by the boron atom is deficient by one electron and its negative electric charge; the boron atom is said to form an electrically positive "hole" in the silicon crystal.

When an electrical force, or voltage, is applied to the boron-doped silicon, it induces an electron from a neighboring atom to leave its position and occupy the electrically positive hole that exists at the site of the boron atom. But as the electron leaves the neighboring atom in order to fill the hole, it leaves that atom with one less electron and so forms another positive hole in the process. The end result of this switching of electrons is that the positive charge that was originally present at the site of the boron atom seems to migrate through the silicon crystal, creating an electrical current. Because the carrier of electricity in this case is a positive charge, the transistor is called a p-type semiconductor. P-type semiconductors are also used in the manufacture of solar cells, in which light generates an electric current.

Although boron is much less reactive than the elements that immediately precede it in the periodic table, it is never found in its pure form in nature.

Carbon

IA																	VIIIA
H	IA											IIIA	IVA	VA	VIA	VIIA	He
Li	Be											B	C	N	O	F	Ne
Na	Mg	IIIB	IVB	VB	VIB	VIIB	VIIIB			IB	IIB	Al	Si	P	S	Cl	Ar
K	Ca	Sc	Ti	V	Cr	Mn	Fe	Co	Ni	Cu	Zn	Ga	Ge	As	Se	Br	Kr
Rb	Sr	Y	Zr	Nb	Mo	Tc	Ru	Rh	Pd	Ag	Cd	In	Sn	Sb	Te	I	Xe
Cs	Ba	*La	Hf	Ta	W	Re	Os	Ir	Pt	Au	Hg	Tl	Pb	Bi	Po	At	Rn
Fr	Ra	†Ac	Rf	Db	Sg	Bh	Hs	Mt	Ds	Rg	Cn	Uut	Uuq	Uup	Uuh	Uus	Uuo

*	Ce	Pr	Nd	Pm	Sm	Eu	Gd	Tb	Dy	Ho	Er	Tm	Yb	Lu
†	Th	Pa	U	Np	Pu	Am	Cm	Bk	Cf	Es	Fm	Md	No	Lr

Atomic Number 6

Chemical Symbol C

Group IVA

Carbon represents only 0.09 percent of the Earth's crust by mass, yet it is the element most essential for life on our planet. If we examine the molecules that make up plants and animals, almost all of them contain carbon. A whole branch of chemistry, called organic chemistry, is essentially the chemistry of carbon compounds. Among the more than 5 million compounds that are considered organic compounds are hydrocarbons, or coal and petroleum products, as well as perfumes, proteins, benzene, enzymes, carbohydrates, and a host of others too numerous to mention. Carbon owes its central position in the organic world to the ability of its atoms to link up with other carbon atoms to form long chains that are either straight or branched. These chains act as a backbone to which other elements are attached, yielding an almost endless number of possible carbon-containing molecules. One such long-chain carbon molecule is deoxyribonucleic acid, or DNA, found in the genetic material of all living creatures. DNA stores information for the construction of an immense number of proteins manufactured by plant and animal cells, and it can replicate itself, or produce copies of itself. Found in the nucleus of plant and animal cells, it is essential to the reproduction of living organisms.

Besides its occurrence in organic compounds, carbon also exists in its free state in nature, where it is found in several different forms. When an element can exist in several natural forms, these forms are called allotropes. In the case of carbon, the most spectacular of these allotropes is diamond, the hardest of materials. However, carbon is also found in the forms of graphite and coal. The structure of graphite consists of sheets of carbon atoms bonded together to form a hexagonal pattern similar to ordinary chicken fencing. The sheets, a single atom thick, are called graphene and are arranged like a deck of playing cards with one sheet on top of another. Like cards, the graphene sheets can slide very easily past each other. The trace left when writing with a pencil, for example, consists of graphene sheets being stripped from the graphite tip of the pencil.

Although diamonds—one of the hardest natural substances known—are prized for their beauty, their main use is industrial. Their hardness makes them excellent cutting tools.

A new allotropic form of carbon, created in 1985, has caused quite a bit of excitement in the world of chemistry. This new molecule contains 60 carbon atoms arranged in a giant cage-like structure that looks very much like a soccer ball. Sir Harold W. Kroto at the University of Sussex in England and Robert F. Curl Jr. and Richard E. Smalley at Rice University in Texas were awarded the Nobel Prize in Chemistry in 1996 for the discovery of the new C(60) molecule. They used a laser beam to vaporize graphite and let it condense in an atmosphere of helium. They named the molecule buckminsterfullereen after the American architect Buckminster Fuller, who was famous for designing geodesic domes, structures that also resembled a soccer ball. Chemists impatient with long names have begun calling these unusual molecules "buckyballs." Variations of these molecules containing as many as 70 carbon atoms have been identified, and there is evidence that many more such clusters exist. The growing family of these compounds is called the fullerenes. There are indications that some fullerenes may be excellent conductors of electricity and may even become superconductors at temperatures close to room temperature. A superconducting material has no resistance whatsoever to the flow of electricity and is an immensely efficient and highly economical conductor of electric current. Most materials become superconductors only at temperatures that are hundreds of degrees below zero Celsius.

It is hard to imagine that graphite, the material found in every pencil, is made of the same material as diamond. But synthetic diamonds can be made from graphite by applying very high pressures and temperatures to it. The synthetic diamonds made in this way are not considered as beautiful as natural diamonds, but they are just as hard and are used in industry for their abrasive properties. The reverse transformation is also possible and in fact occurs spontaneously. Luckily, the process is a very slow one and can take several million years, but eventually diamonds will disintegrate.

The study of graphite led to the discovery of yet another allotropic form of carbon in 1991. The Japanese physicist Sumio Iijima, studying the surface of graphite rods that were used in electric arc discharges, accidentally discovered that graphene could be rolled up like a section of fencing to form a tiny closed cylinder, called a nanotube. This tube with a diameter that typically has an order of magnitude of a few nanometers (a nanometer is 1 billionth of a meter) is thousands of times smaller than a human hair. Often found as single tubes, nanotubes sometimes contain even smaller tubes inside them, forming what are called multiwalled tubes.

Although their widths are tiny, nanotubes can be fabricated with lengths that are thousands of times longer than their diameter. These fibers, or ropes, as they are sometimes called, are among the strongest materials known, with tensile strengths that are 100 times that of steel. New applications for nanotubes are constantly being announced in such fields as electronics, medicine, and solar energy, and their properties are being extensively studied in laboratories throughout the world.

Two Russian scientists, Konstantin Novoselov and André Geim of the University of Manchester, in England, were awarded the 2010 Nobel Prize in Physics for studying the properties of a sheet of graphene, 6 years earlier. In a groundbreaking paper published in October 2004, they described how they managed to isolate and identify a single sheet of graphene using Scotch tape to peel off a layer from a sample of graphite—a layer that remained flat and that did not curl up into a buckyball or nanotube.

This graphene sheet made of carbon, one atom thick, the "first two-dimensional crystal," as they called it, has remarkable properties. It is basically the thinnest material known, is transparent, is an excellent conductor of electricity, conducts heat better than any known material, and, like nanotubes, has a tensile strength stronger than steel. Many scientists think that its potential for applications in the world of computers and electronics is unlimited.

Coal, a mixture of carbon and various other compounds, contains carbon in a noncrystalline form. It is a major source of energy for industry and domestic use. Like oil and natural gas, coal is considered a fossil fuel because it results from the decay of plants and animals over hundreds of millions of years. Hard or anthracite coal is the older variety of coal and is thought to be about 250 million years old. It contains about 80 percent carbon. Bituminous or soft coal contains only about 40 to 50 percent carbon and is a younger variety of coal. Data from the U.S. Department of Energy indicate that as of 1991, 23 percent of the energy used in the United States came from coal, approximately 40 percent from petroleum, and 24 percent from natural gas. Fossil fuels thus contribute about 90 percent of the total energy consumed in the United States.

When coal and other fossil fuels are burned, carbon dioxide (CO_2), which contributes to global warming, is released into the air.

Coke, which is used in great quantities in producing iron and steel, is produced by heating coal in the absence of air. In this process, almost all of the volatile substances and impurities in coal are burned off, leaving a fairly pure form of carbon. Coke or charcoal has been known since ancient times, and the name carbon is in fact derived from the Latin *carbo*, for "charcoal." A coke or charcoal fire is hotter than an ordinary wood fire and was of prime importance in the development of ancient techniques for recovering iron from its ore.

There are many problems associated with the use of coal and other fossil fuels as sources of energy. One problem is that many of these fuels also serve as raw materials for the manufacture of a host of useful products such as plastics and medicines. At the rate at which we are using these materials as fuels, it has been estimated that our petroleum and natural gas supplies will be completely depleted by the year 2030. Of more immediate concern, however, is that the products of carbon combustion represent a major source of pollution. They are responsible for acid rain and include large amounts of carbon dioxide, or CO_2, a colorless, odorless gas that is formed when carbon is burned in air and that contributes to the possibility of global warming.

Besides being formed by the combustion of carbon in air, carbon dioxide is also a by-product of metabolism in animals, all of whom exhale this gas as they breathe. Moreover, when animals and plants die, the process of their decomposition produces carbon dioxide. Volcanoes are usually not thought of as agents of pollution, but they are also a major source of carbon dioxide. Given all of these sources of the gas, it is fortunate that plants use carbon dioxide to produce carbohydrates, through the process of photosynthesis. Until recently, the production and consumption of carbon dioxide resulted in a dynamic equilibrium, so that the concentration of this gas in the atmosphere was more or less constant. There seems to be evidence, however, that the rapid growth of industrial society with its excessive use of fossil fuels has led to a steadily increasing concentration of carbon dioxide in the atmosphere.

Carbon represents only 0.09 percent of the Earth's crust by mass, yet it is the element most essential for life on our planet.

Carbon dioxide influences the temperature on Earth by what is usually called the "greenhouse effect." Like the glass roof of a greenhouse, carbon dioxide in the atmosphere permits most of the energy radiated by the sun to reach the Earth. And like the glass of a greenhouse, carbon dioxide also absorbs the infrared radiation given off by the heated Earth, effectively trapping the heat generated by this radiation and thus warming the Earth. Many dire predictions have been made about the catastrophic consequences of allowing the buildup of carbon dioxide in the atmosphere to continue. Some environmentalists fear that the melting of icecaps and glaciers, from the warming produced by the gas, will cause the level of the oceans to rise and flood coastal areas. It is also feared that vast global changes in climate from the warming process could produce deserts in areas now fertile. A worldwide alert is in effect, and many advanced industrial countries are taking measures to reduce the levels of carbon dioxide and other substances emitted by the burning of fossil fuels.

Despite all of the ominous problems associated with the buildup of carbon dioxide in the Earth's atmosphere, it does have many beneficial uses. Commercially, carbon dioxide is obtained as a by-product in the manufacture of ammonia. Whenever you drink soda water, you are drinking water that has been carbonated by having carbon dioxide dissolved in it. All of the natural bodies of water on Earth also contain dissolved carbon dioxide. In solution it forms a weak acid called carbonic acid. Structurally, carbonic acid consists of two hydrogen ions, each of which is designated as H^+ because of its single positive charge, and a carbonate group, $(CO_3)^=$, which carries a double negative charge and is easily dissociated from the H^+ ions that give the acid its acidic properties. Because of this, it is not surprising that there are carbonate deposits in areas formerly covered by water. Carbonate is left in the Earth's crust chiefly as calcite, limestone, dolomite, marble, and chalk. Often, stalactites and stalagmites are formed in caves by the crystallization of carbonates that were dissolved in groundwater.

When carbon dioxide is frozen, it is called dry ice and is used as a refrigerant.

When carbon dioxide is frozen, it is called dry ice and is used as a refrigerant. Because carbon dioxide freezes at −78°C, dry ice is considerably colder than ice made from water. Another property that makes dry ice an excellent refrigerant is its ability to pass directly from the solid state to the gaseous state as it warms, without first becoming a liquid. Chemists call this process sublimation. When steaks are shipped from Omaha to New York by mail, for example, dry ice is the refrigerant of choice for preserving them because it eliminates the problem of having to deal with the liquid that would form from the melting of ordinary ice.

Nevertheless, it is possible to melt dry ice by heating it at high pressures. Under these conditions, the transition from solid to liquid takes place at a temperature of about −56°C. Liquid carbon dioxide is used as a solvent in extracting caffeine from coffee to make decaffeinated coffee. Its great advantage is that it leaves no residue in the coffee.

Graphite tiles line the inner wall of the Tokamak Fusion Test Reactor at the Princeton Plasma Physics Laboratory.

Carbon dioxide is a very heavy gas, with a density about 1.5 times greater than that of air. It tends to collect close to the ground in unventilated spaces, and by displacing the air it can be hazardous to humans. Every year, a number of workers are killed by carbon dioxide asphyxiation while cleaning out enclosed storage tanks aboard ships. At the same time, its property of excluding air makes carbon dioxide useful in fighting fires. Fire extinguishers often contain compressed carbon dioxide that is released to smother flames by displacing the air on which they feed, making combustion impossible.

When carbon is burned in the absence of sufficient oxygen, it forms a gas called carbon monoxide, CO. Industry uses carbon monoxide chiefly as a fuel, although it is also used in the metallurgical industry to help recover metals from their oxide ores. Unlike carbon dioxide, carbon monoxide is a very poisonous gas. It combines with the hemoglobin in blood, preventing the latter from carrying its usual load of oxygen to the tissues of the body. Even a small amount of carbon monoxide can cause drowsiness and severe headaches. An improperly vented garage or gas heater can produce carbon monoxide in large enough quantities to cause asphyxiation and death. Carbon monoxide is also produced in automobile exhaust and has become a major source of pollution in large cities.

The form of carbon known as activated charcoal consists of a very finely powdered charcoal. It has an enormous surface area, which may reach 1,000 square meters per gram. This surface is very effective in adsorbing other molecules, or attracting them and holding them on the surface of the charcoal granule. Activated charcoal is therefore used extensively in devices designed to remove pollutants from the air.

When natural gas, a carbon-containing gas left in the Earth in the form of deposits from decaying plant and animal matter, is burned in a special oven with little air, it produces a powdered carbon called carbon black or lampblack. Lampblack has been known and used for thousands of years. The ancient Egyptians used lampblack to make an ink that could be used for writing on papyrus. It is still used by the printing industry today. Lampblack is also used in the manufacture of automobile tires to increase their durability, and with the addition of proper adhesives it can easily be molded into various shapes for special purposes. The cylinders of carbon that form the electrodes in a dry cell, for example, are made from lampblack.

When silicon is heated with carbon to a fairly high temperature, it forms a compound called silicon carbide, more commonly known as carborundum. Carborundum is almost as hard as diamond and is used as an abrasive for polishing glass and metals.

The combination of carbon and nitrogen produces a class of organic chemical compounds called cyanides. Hydrogen cyanide, which has a characteristic aroma resembling that of bitter almonds, is a deadly poison and is the gas used in execution chambers. Even 0.1 or 0.2 of 1 percent by volume of hydrogen cyanide in air can be fatal. Hydrogen cyanide is toxic because it interferes with the normal workings of iron-containing molecules in the body. By disabling a key enzyme used in metabolism, it can cause asphyxiation and death within minutes.

The isotope of carbon known as carbon-14 has become a very useful tool for dating relics and archaeological artifacts. The method for doing this was first developed by Willard F. Libby, an American chemist at the University of Chicago, who was awarded the Nobel Prize in chemistry in 1960 for this work.

The isotope of carbon known as carbon-14 —with a half-life of 5,730 years—has become a very useful tool for dating relics and archaeological artifacts.

Carbon-14 is radioactive, with a half-life of 5,730 years. It is constantly being made in the upper layers of the Earth's atmosphere by high-energy particles from outer space that interact with the nuclei of nitrogen molecules present in the atmosphere. This rain of energetic particles—called cosmic rays—has always been part of the background radiation to which all inhabitants of the Earth are subject.

The radioactive carbon-14 generated by this process behaves chemically like ordinary carbon and combines with oxygen to form carbon dioxide. This carbon dioxide is, however, radioactive, and when it is taken up by plants in photosynthesis it produces slightly radioactive living tissue. Animals feeding on these plants then take up the radioactive isotope and eliminate it as natural waste material. Experiments have shown that as a result of this cycle all living things maintain a fairly steady ratio of radioactive carbon to normal carbon. Once a plant or animal dies, however, the natural loss of carbon-14 as a result of radioactive decay continues without any compensating intake of the isotope. When this happens, the ratio of the quantity of C-14 to that of ordinary carbon in the residue of the plant or animal acts as a clock, decreasing at a rate determined by the half-life of the radioisotope, and can be used to determine when the plant or animal died. This technique has been used to date ashes from ancient campfires, parchments, bones from relics, and ancient fabrics. A fabric that made headlines recently was the famous Shroud of Turin, a material that supposedly dated from the time of Christ. Carbon dating at several laboratories established, however, that the shroud was less than 600 years old.

IA	IA	IIIB	IVB	VB	VIB	VIIB	VIIIB	VIIIB	VIIIB	IB	IIB	IIIA	IVA	VA	VIA	VIIA	VIIIA
H																	He
Li	Be											B	C	N	O	F	Ne
Na	Mg											Al	Si	P	S	Cl	Ar
K	Ca	Sc	Ti	V	Cr	Mn	Fe	Co	Ni	Cu	Zn	Ga	Ge	As	Se	Br	Kr
Rb	Sr	Y	Zr	Nb	Mo	Tc	Ru	Rh	Pd	Ag	Cd	In	Sn	Sb	Te	I	Xe
Cs	Ba	*La	Hf	Ta	W	Re	Os	Ir	Pt	Au	Hg	Tl	Pb	Bi	Po	At	Rn
Fr	Ra	†Ac	Rf	Db	Sg	Bh	Hs	Mt	Ds	Rg	Cn	Uut	Uuq	Uup	Uuh	Uus	Uuo

*	Ce	Pr	Nd	Pm	Sm	Eu	Gd	Tb	Dy	Ho	Er	Tm	Yb	Lu
†	Th	Pa	U	Np	Pu	Am	Cm	Bk	Cf	Es	Fm	Md	No	Lr

Nitrogen

Atomic Number **7**

Chemical Symbol **N**

Group **VA**

Nitrogen heads the family of elements that make up Group VA in the periodic table. Nitrogen is a gas, whereas all the other elements that make up the family in the group are either metals or are metal-like. As a gas, nitrogen is relatively inert and is without color, taste, or odor. We are constantly breathing in large quantities of nitrogen as we inhale air, but its lack of any sense-stimulating property makes its presence unnoticed.

Nitrogen dominates the gases in the Earth's atmosphere, making up some 78 percent of the air by volume. It is striking to note that the volume of nitrogen in the air is about four times that of oxygen. Although a relatively unreactive gas, nitrogen forms hundreds of thousands of compounds that are of crucial importance for agriculture and industry. Considering the stability of nitrogen itself, some of these compounds are, ironically, extremely unstable and explosive.

Nitrogen was discovered in 1772 by Daniel Rutherford (1749–1819), an English chemist and physician. Using a bell jar to confine a sample of air, he first removed all the oxygen from the trapped air by burning a substance in it. By placing an unfortunate mouse inside the jar and demonstrating that it asphyxiated, he then showed that the residual gas in the jar, later named nitrogen, could not support life.

Approximately 30 million tons of gaseous nitrogen are produced each year from liquefied air. In this process, called fractional distillation, air is first cooled until it liquefies and then gradually warmed. Each of the gases that make up the air has a different boiling point and boils off separately when this is done, so that it is relatively easy to separate the nitrogen, which has a

The combination of nitrogen with oxygen in the internal combustion engines of automobiles produces nitrogen oxide in a reaction similar to the one caused by lightning.

boiling point of −195.8°C, from the remaining mixture and collect it. Liquid nitrogen is used industrially to freeze foods and to preserve biological specimens. Semen from a bull, for example, can be kept immersed in liquid nitrogen for long periods before being used in artificial insemination. Because it is nonreactive, liquid nitrogen is also the refrigerant of choice in experiments requiring very cold conditions, such as the testing of superconducting electrical materials.

Nitrogen in its gaseous form is often used in situations in which it is important to keep other, more reactive atmospheric gases away. It serves industry as a blanketing gas, for example, in protecting materials such as electronic components during production or storage. To prevent the oxidation of wine, wine bottles are often filled with nitrogen after the cork is removed. Nitrogen has recently also been used in blanketing fruit after it has been picked to protect it from deterioration. Apples, for example, can be stored for up to 30 months if they are kept at low temperatures in an atmosphere of nitrogen. In addition to these applications, nitrogen is used in oil production, in which it is pumped in compressed form underground to force oil to the surface. The method is called enhanced oil production. Ordinary air cannot be used for this purpose because some of the gases that make up air would react with the oil, producing undesired by-products.

Many compounds that contain nitrogen are crucial for plant and animal life. Among the great variety of biological molecules that have nitrogen as a component are proteins and nucleic acids. Despite the lack of reactivity of nitrogen, nature has developed several mechanisms for converting nitrogen for use by living cells. The process of converting nitrogen from the atmosphere into usable nitrogen compounds is called nitrogen "fixing."

Some nitrogen is fixed by lightning. During electrical storms, the extremely high temperatures produced near a bolt of lightning supply enough energy to break apart the normally diatomic nitrogen molecule. The free nitrogen atoms that are formed can then combine with oxygen to form nitrogen oxide, NO, and nitrogen dioxide, NO_2. Most of the nitrogen dioxide dissolves in rainwater and then falls to the Earth's surface, where nitrogen-fixing bacteria in the soil called cyanobacteria use the nitrogen dioxide to build nutrients such as proteins and amino acids. These nutrients are then taken up by the roots of plants, which are in turn eaten by animals, in whose tissues the nitrogen is used to create animal proteins. Bacteria in the soil then convert the inevitable waste products of nitrogen metabolism, such as urea, into amino acids and ammonia. The same denitrifying bacteria also break down these compounds to form gaseous nitrogen, which is returned to the atmosphere.

The complex natural system of reactions that leads from nitrogen fixing in the soil to the eventual return of nitrogen to the atmosphere is called the "nitrogen cycle." Another natural method of nitrogen fixation uses a class of plants called legumes. These plants, which include soybeans, alfalfa, and clover, have root nodules that contain nitrogen-fixing bacteria. These bacteria generate an enzyme called nitrogenase that converts nitrogen trapped in the soil directly into ammonia. How this is done is still a mystery, but these nitrogen-fixing plants are often planted in rotation with other food crops to restore the biological vigor of the soil.

Unfortunately, modern farming is very intensive and often depletes the soil of vital nitrogen compounds. In order to feed a rapidly growing population, artificial fertilizers are required. Ammonia is the compound most often used as a source of nitrogen for these fertilizers. Ammonia is probably the most important commercial compound of nitrogen. It is a colorless gas with a characteristic pungent odor that many people find irritating. The ammonia one buys in a grocery store is actually a solution of ammonia in water. The name for ammonia has an interesting derivation. A compound of ammonia that we now call ammonium chloride was originally prepared from animal wastes in ancient Egypt near a temple dedicated to the god Ammon. After being brought to Europe, it assumed the name of sal ammoniac, or the salt of Ammon.

Almost all of the ammonia produced commercially today is made using the Haber process. In 1905 the German chemist Fritz Haber (1868–1934) accomplished what chemists had thought virtually impossible. He showed that it was possible to combine nitrogen and hydrogen directly to produce ammonia. The reaction required a temperature of about 500°C, a pressure as high as 1,000 times normal atmospheric pressure, and an iron catalyst. Haber was awarded the Nobel Prize in 1918 for this pioneering work.

The Haber process proved to be of crucial importance for Germany during World War I, when that country was cut off from

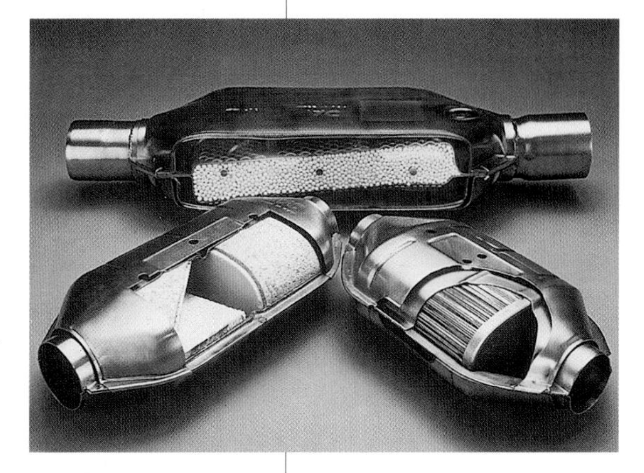

The use of catalytic converters in automobiles has reduced the amount of harmful nitrogen oxides released into the atmosphere.

many essential raw materials by an Allied naval blockade. Not only was the process used to make crucial fertilizers but also the ammonia generated by the Haber process was used to make nitric acid, a chemical essential for the manufacture of explosives.

Nitrogen oxides, formed by the combination of nitrogen with oxygen, have contributed to the pollution that has plagued many cities. The brownish haze that one often sees hovering over some cities is caused by the presence of nitrogen dioxide.

Some of the nitrogen oxides, formed by the combination of nitrogen with oxygen, have contributed to the pollution that has plagued many cities. The combination of nitrogen with oxygen in the internal combustion engines of automobiles produces nitrogen oxide, or NO, in a reaction similar to the one caused by lightning. When released into the air, this NO reacts with more oxygen to form nitrogen dioxide, NO_2, which is an extremely corrosive gas. The brownish haze that one often sees hovering over such cities as Los Angeles is caused by the presence of nitrogen dioxide. Besides being corrosive, NO_2 is quite noxious and can cause considerable damage to plants and animals. To make matters worse, the ultraviolet radiation present in sunlight can dissociate, or break apart, nitrogen dioxide molecules to produce free oxygen atoms. These so-called *oxygen radicals* are extremely reactive and combine with oxygen molecules to form ozone, O_3. Ozone is a powerful oxidizing agent that is harmful to plants and animals as well as many structural materials such as rubber and plastic. It also reacts with the exhaust from automobiles to form a range of organic pollutants that are very irritating to the eyes and throat. Photochemical smog is the name usually given to the pollutants produced from automobile exhaust in the presence of sunlight. The word *smog* was originally used to describe the noxious combination of smoke and fog that occurred regularly in London in the 1950s.

Nitrogen dioxide in the air can also react with water in the air to form a corrosive acid known as nitric acid. In small quantities this can be beneficial, contributing nitrogen to the soil. In large quantities, however, the nitric acid droplets that fall to earth during rainstorms cause considerable damage to buildings, monuments, and animal life.

One of the chief methods of reducing the emission of nitrogen oxides has been through the use of catalytic converters. The catalytic converter built into the exhaust system of today's automobiles uses a mixture of powdered catalysts to decompose the nitrogen oxide in automobile exhaust into harmless nitrogen and oxygen.

A third oxide of nitrogen, nitrous oxide, or N_2O, has properties quite different from those of the two compounds described above. It is better known as "laughing gas," because a person inhaling this gas usually becomes lightheaded and somewhat intoxicated. Nitrous oxide is a stable gas with a slightly sweet odor that is often used in dentistry as a mild anesthetic. The gas also dissolves in cream under pressure, making it useful as a propellant in whipped-cream dispensers. When the pressure is removed by dispensing the cream, the gas bubbles out of solution, forming a whipped-cream foam.

Nitric acid, or HNO_3, is of major commercial importance in the United States, where it is used to produce the fertilizer called ammonium nitrate, as well as for the manufacture of explosives and such plastics as nylon and polyurethane. Nitric acid was known to early chemists as *aqua fortis*, which in Latin means "strong water." More than 15 billion pounds of nitric acid were produced in the United States in 1992, chiefly by the Ostwald process. This process was discovered in 1902 by the German chemist Wilhelm Ostwald (1853–1932), who received the Nobel Prize in 1909 for his work on the importance of catalysts in chemical reactions. In the Ostwald process, high temperatures and platinum catalysts convert ammonia into nitric acid. Because nitric acid is used to make explosives, Ostwald's work allowed Germany to continue fighting in World War I after the Allies cut off all supplies of nitrates from Chile.

When nitric acid reacts with glycerol, a viscous type of alcohol, it creates an explosive compound of nitrogen called nitroglycerine. It is extremely dangerous to work with nitroglycerine because it can explode at the slightest shock, releasing large quantities of heat and large volumes of nitrogen and carbon dioxide gas. It is the rapid expansion of these gases that causes the shock wave associated with the explosion.

A useful derivative of nitroglycerine, called dynamite, was discovered in 1867 by Alfred Nobel (1833–1896). He showed that when nitroglycerine is mixed with clay, it forms a relatively safe, shock-resistant explosive. Nobel later established the Nobel Prize with the fortune he made from this discovery.

When nitric acid is neutralized with bases such as sodium and potassium hydroxide, it forms compounds called nitrates. Nitrates and another group of nitrogen oxide compounds known as nitrites are added to canned and other preserved food to help prevent the growth of botulism bacteria and to preserve the appearance of meat products. After some time, stored meat begins to turn brown because of the oxidation of blood in the meat. Both nitrates and nitrites slow this process so that the meat becomes less objectionable to consumers. Many

Sodium azide (NaN_3), an explosive compound of nitrogen, is used in the airbags found in cars today.

consumer groups have objected to the use of these compounds, however, because there seems to be evidence that when eaten and subjected to the acidic conditions of the stomach, they give rise to a carcinogenic, or cancer-causing, group of compounds called nitrosamines.

One of the isotopes of nitrogen, nitrogen-13, is used with the new medical technology known as positron emission tomography, or PET scanning. Nitrogen-13 is radioactive, giving off subatomic particles called positrons when it decays. Positrons are really positive electrons that interact with ordinary electrons to produce radiation resembling X rays. When a positron emitter like nitrogen-13 is injected into the body, a special scanner around the body records this radiation and is capable of producing a cross-sectional picture of the body, including internal organs. Because nitrogen-13 has a short half-life (9.97 minutes), a dose of this isotope that is administered to a patient is quickly dissipated by natural decay before it can cause any radiation damage. PET scans have proven particularly valuable in helping to diagnose diseases involving brain dysfunction, such as schizophrenia and Alzheimer's disease.

Nitrogen dominates the gases in the Earth's atmosphere, making up some 78 percent of the air by volume.

An interesting compound of nitrogen that has recently found widespread use is sodium azide, NaN_3. This colorless salt is used in the airbags found in almost every automobile today. Sodium azide is very explosive and on impact or ignition it will very quickly decompose to form large volumes of nitrogen gas. The gas liberated by the explosion inflates the bag and provides a cushion to soften the effects of a collision.

In 1999, a team of chemists headed by Karl O. Christie and William W. Wilson at the Air Force Research Laboratory at the Edwards Air Force Base in California discovered a new form of nitrogen. The new compound is a molecule that consists of five nitrogen atoms bound to each other in the shape of a V.

The discovery surprised most experts, who felt that any molecule containing more than three nitrogen atoms would require so much energy and be so unstable that it would be impossible to fabricate in the laboratory. The new substance, which had the five nitrogen atoms bound to an ion of arsenic and fluorine, was indeed so sensitive that the few grains synthesized exploded, destroying most of the equipment being used to analyze it. More work on this new allotropic form of nitrogen is nevertheless continuing in the search for possible uses.

Oxygen

IA	IA	IIIB	IVB	VB	VIB	VIIB	VIIIB			IB	IIB	IIIA	IVA	VA	VIA	VIIA	VIIIA
H																	He
Li	Be											B	C	N	O	F	Ne
Na	Mg											Al	Si	P	S	Cl	Ar
K	Ca	Sc	Ti	V	Cr	Mn	Fe	Co	Ni	Cu	Zn	Ga	Ge	As	Se	Br	Kr
Rb	Sr	Y	Zr	Nb	Mo	Tc	Ru	Rh	Pd	Ag	Cd	In	Sn	Sb	Te	I	Xe
Cs	Ba	*La	Hf	Ta	W	Re	Os	Ir	Pt	Au	Hg	Tl	Pb	Bi	Po	At	Rn
Fr	Ra	†Ac	Rf	Db	Sg	Bh	Hs	Mt	Ds	Rg	Cn	Uut	Uuq	Uup	Uuh	Uus	Uuo

*	Ce	Pr	Nd	Pm	Sm	Eu	Gd	Tb	Dy	Ho	Er	Tm	Yb	Lu
†	Th	Pa	U	Np	Pu	Am	Cm	Bk	Cf	Es	Fm	Md	No	Lr

Atomic Number 8

Chemical Symbol O

Group VIA

Oxygen is one of the most important and abundant elements on Earth. It exists in the atmosphere as a gas, in water as part of the water molecule, and in the Earth's crust in an enormous variety of rocks and minerals. Some 46 percent of the Earth's crust is oxygen, and it is by far the most abundant element found there. It is essential for life and is part of almost every biological molecule in our bodies. The combustion of various fuels depends upon oxygen and provides most of the energy for industry and for heating.

In its most common elemental form, oxygen exists in the atmosphere as a diatomic molecule, O_2. It is colorless and odorless

Through the process of photosynthesis, plant cells containing chlorophyll—a green pigment present in most plants—use energy from the sun to convert carbon dioxide into carbohydrates and oxygen.

Oxygen is by far the most abundant element in the Earth's crust. It is essential for life and is part of almost every biological molecule in our bodies.

and makes up about 20.95 percent of the atmosphere by volume. Although many natural processes consume oxygen, such as combustion and the decay of organic matter, this gas is constantly being replenished by photosynthesis in plants. In photosynthesis, the chlorophyll in plants uses energy from the sun to convert carbon dioxide into complex carbohydrates and oxygen. Oxygen is then released into the atmosphere. It is thus continually being consumed and continually being produced in a cycle that has resulted in an almost constant amount of oxygen in the Earth's atmosphere.

Credit for the discovery of oxygen is usually given to Joseph Priestley (1733–1804), an English chemist. In a famous experiment he heated an oxide of mercury and noted that the gas it gave off caused a candle "to burn with a remarkably brilliant flame." The gas was oxygen. Credit for naming the element is usually given to the French chemist Antoine Laurent Lavoisier (1743–1794). The name is derived from the Greek words *oxys*, which means acid, and *gens*, which means creator or former.

We depend on oxygen to live, of course, because it is essential for so many of the biological processes that take place in our bodies. Oxygen is transported from our lungs to the cells in our bodies by means of a large protein molecule called hemoglobin. This molecule is enormous, being made up of some 574 chemical units called amino acids. Located in our red blood cells, the hemoglobin chemically binds oxygen to itself and then gives it up to our tissues. When charged with oxygen, the hemoglobin is bright red and gives our blood its characteristic color. When the oxygen is released, the color changes and becomes bluish.

Normal red blood cells are shaped like little discs. In some people, however, a few of the amino acids that make up the hemoglobin molecule are faulty. This causes a dramatic change in the shape of the molecule and the red blood cell. The cell now becomes sickle-shaped rather than disc-like. The change in shape causes serious problems in the transfer of oxygen and produces a condition known as sickle cell anemia.

Oxygen also exists as a triatomic molecule called ozone, the chemical formula of which is O_3. Unlike ordinary oxygen, ozone has a faintly blue color and a characteristic brackish odor. It can be created by passing electrical discharges through oxygen and is therefore very noticeable near high-voltage electrical motors in subway and railroad stations and during electrical storms.

Ozone is very reactive and is quite destructive to materials such as rubber and fabrics. It is also quite harmful to lung tissue, and during periods when there is an excess amount of ozone in the air it is usually suggested that older people and children not engage in any strenuous physical activity that would increase the deep inhalation of ozone. State and local authorities carefully monitor the amount of ozone present in the air and compare it to the allowable amounts suggested in the U.S. National Ambient Air Quality Standard. The suggested maximum daily 1-hour concentration of ozone at any location is 120 parts per million parts of air, an amount that is often exceeded in congested cities. The chief source of ozone in the lower atmosphere, sometimes called "bad ozone," is the photochemical destruction of nitrogen dioxide in the exhaust of automobiles. This process has already been described in the section on nitrogen.

In contrast to bad ozone, "good ozone" exists in the upper layers of the Earth's atmosphere. This ozone shields the surface of the Earth from the ultraviolet radiation emitted by the sun, which would otherwise be strong enough to destroy living tissue.

Oxygen combines with hydrogen to form water, or H_2O, which is one of the most common molecules on Earth.

Because oxygen combines with almost every element, the compounds it forms are too numerous to describe. The most common oxygen compounds in the Earth's crust are oxides. Some examples of these solid oxides, in comparison with gaseous oxides such as carbon dioxides and nitrogen oxides, are silicon dioxide, calcium oxide, aluminum oxide, and magnesium oxide. Hematite, or ferric oxide, is a common ore from which iron is extracted.

Oxygen combines with hydrogen to form water, or H_2O, which is one of the most common molecules on Earth. Water is unusual in that its density as a liquid is greater than its density as a solid. This explains why ice cubes float in water and why ice forms on the surface of a freezing lake. If water did not have this physical property, lakes would freeze from the bottom up in winter, and no life would be possible in such bodies of water. Oxygen can also combine with hydrogen to form a liquid called hydrogen peroxide, H_2O_2, whose properties are quite different from those of water. Hydrogen peroxide is used chiefly as an industrial and cosmetic bleach and as a disinfectant.

Large quantities of pure oxygen for industrial and aerospace use are usually recovered from cooled liquid air by first boiling off the other major constituents of air, nitrogen and argon. The oxygen can then be transported and used in its cooled liquid state or stored as a gas under pressure. Small quantities of oxygen, however, are often supplied by heating sodium or potassium chlorate, usually in the presence of a catalyst to speed up the reaction. These compounds decompose to produce oxygen. On an airplane, for example, small canisters containing sodium chlorate and iron are placed above every seat. If for some reason oxygen is needed, a trigger-like device sets off a small explosion, mixing the two chemicals and producing oxygen.

Fluorine

IA	IA	IIIB	IVB	VB	VIB	VIIB	VIIIB	VIIIB	VIIIB	IB	IIB	IIIA	IVA	VA	VIA	VIIA	VIIIA
H																	He
Li	Be											B	C	N	O	F	Ne
Na	Mg											Al	Si	P	S	Cl	Ar
K	Ca	Sc	Ti	V	Cr	Mn	Fe	Co	Ni	Cu	Zn	Ga	Ge	As	Se	Br	Kr
Rb	Sr	Y	Zr	Nb	Mo	Tc	Ru	Rh	Pd	Ag	Cd	In	Sn	Sb	Te	I	Xe
Cs	Ba	*La	Hf	Ta	W	Re	Os	Ir	Pt	Au	Hg	Tl	Pb	Bi	Po	At	Rn
Fr	Ra	†Ac	Rf	Db	Sg	Bh	Hs	Mt	Ds	Rg	Cn	Uut	Uuq	Uup	Uuh	Uus	Uuo

*	Ce	Pr	Nd	Pm	Sm	Eu	Gd	Tb	Dy	Ho	Er	Tm	Yb	Lu
†	Th	Pa	U	Np	Pu	Am	Cm	Bk	Cf	Es	Fm	Md	No	Lr

Atomic Number 9

Chemical Symbol F

Group VIIA—
The Halogens

Fluorine heads the column of elements in the periodic table known as the halogens, or "salt formers." The other members of this family are chlorine, bromine, iodine, and astatine. Astatine, at the bottom of the group, is highly radioactive. All halogen atoms react chemically by accepting electrons from other atoms and readily combine with metals to form compounds known as salts. The smaller the halogen atom, the closer to the nucleus the incoming electron will be and the greater the force that attracts it and makes the reaction occur. Fluorine is the smallest, lightest, and most reactive member of the halogen group.

Fluorine is a pale yellow gas whose molecules contain two atoms. It is, however, far too reactive to be found in nature in its diatomic form. Fairly large quantities of fluorine occur in the minerals fluorite or calcium fluoride (sometimes called fluorspar) and cryolite, which contains fluorine, aluminum, and sodium. Fluorine is also found in small quantities in seawater, teeth, bones, and blood.

The French chemist Henri Moissan won the Nobel Prize in 1906 for first producing fluorine in its pure elemental form. Its name is taken from the Latin word *fluere*, which means "to flow." The origin of the name probably derives from the use of fluorite as a flux. A flux is added to many compounds and minerals to lower their melting point and make them more available for further chemical treatment. It was long known that fluorine compounds existed, but because of its extreme reactivity, pure fluorine had never been isolated. It took 75 years of continuous effort on the part of many chemists before it was finally isolated by Moissan in 1886.

Fluorine is usually prepared by passing an electric current through molten fluoride salts. In 1986, however, Karl O. Christie, working at the Air Force laboratory at Edwards Air Force Base in California, astounded many chemists by successfully isolating pure fluorine from a fluorine compound using only chemical means. Because of the large amounts of energy required for such a procedure, it had long been believed that such a chemical separation was impossible.

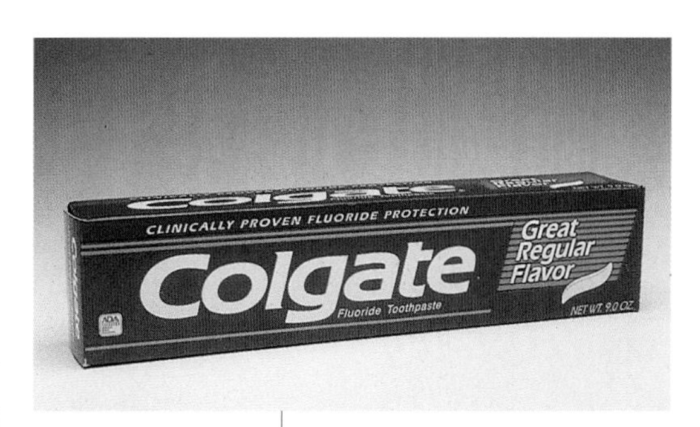

Toothpastes containing fluorine, in the form of sodium fluoride, can help prevent cavities. In the presence of sodium fluoride, a process called remineralization makes a modified form of tooth enamel that is resistant to attack by acids.

Today fluorine is commercially prepared from fluorspar at the rate of thousands of tons a year. It is shipped as a liquid in special containers, cooled by liquid air.

Fluorine gas is used to produce a rather exotic gaseous compound of uranium called uranium hexafluoride (UF_6). It is in the form of this compound that uranium is usually shipped to huge gas diffusion plants, where the important uranium-235 isotope is then separated from the more common uranium-238 isotope. It is the U-235 isotope that is easily fissionable and plays such a major role in nuclear reactors and nuclear weapons.

Another important role for fluorine in nuclear research is in producing a gas for detecting the presence of neutrons. This sensitive gas is a compound of fluorine and boron called boron trifluoride (BF_3). As we have seen in the section on boron, this element readily absorbs neutrons and then emits highly charged alpha particles that are easily detected. However, there are technical obstacles to the use of pure, solid boron for such detectors, and it is far easier to construct a neutron detector that uses the gaseous, boron-containing molecules of BF_3. BF_3 detectors, as they are called, are among the most important tools for this purpose.

Many communities in the United States now add fluorine, in the form of sodium fluoride, to their public water supplies. Research has shown that small quantities of fluorine can retard the occurrence of cavities in teeth. Teeth are protected by a hard enamel that is composed of a mineral called hydroxyapatite. This mineral is not very soluble, but in the presence of the weak acids that bacteria in the mouth produce from food that has a high sugar content, some breakdown of hydroxyapatite occurs. The resulting loss of enamel promotes the onset of tooth decay. One of the body's defenses against the loss of tooth enamel is to make new enamel constantly, a process called remineralization. In the presence of sodium fluoride, the remineralization process makes a modified form of enamel, called fluorapatite, that is more resistant to attack by acid than hydroxyapatite is. Manufacturers of toothpaste now routinely add sodium fluoride to their products.

Considering the extreme reactivity of fluorine gas, it is somewhat surprising that a plastic called Teflon, consisting of long, chainlike molecules of carbon linked chemically to fluorine, should be as inert as it is. Teflon is used for nonstick surfaces on frying pans and on a variety of other products that require smooth, nonreactive surfaces. It is also used to make artificial valves for the heart.

Fluorocarbons in aerosol cans have been banned in the United States because these gases diffuse into the upper atmosphere and the resulting loss of ozone permits dangerous ultraviolet radiation to reach the surface of the Earth.

Other fluorinated compounds of carbon include aerosol propellants for spray cans and Freon, an inert gas used as a refrigerant. The use of fluorinated carbon gases, known as fluorocarbons, in aerosol cans has been banned in the United States since 1978 because it is known that these gases diffuse into the upper atmosphere and react with the ozone layer. The resulting loss of ozone permits dangerous ultraviolet radiation to reach the surface of the Earth.

In the presence of hydrogen, fluorine burns with explosive force. The reaction is spontaneous, and not even a match or spark is needed. The reaction forms hydrogen fluoride, which is an acid when dissolved in water. Hydrofluoric acid is extremely dangerous and must be carefully handled. A small amount on the skin can cause extreme pain. However, this acid has the ability to dissolve glass and is used to etch designs on glass objects.

An artificially made, radioactive isotope of fluorine, fluorine-18, is one of several isotopes used in the medical procedure called positron emission tomography (PET). Fluorine-18 spontaneously emits positrons, which are the antiparticles of electrons and resemble the latter in every respect but their electric charge. When a positron collides with an electron, the two particles "annihilate" each other, yielding a burst of energy that takes the form of X-ray-like radiation. If fluorine-18 is introduced into the body, this annihilation radiation is emitted within the body and can be scanned by special detecting instruments to produce cross-sectional pictures of portions of the body. Fluorine-18 is ideal for this technique because it has a half-life of only 109.8 minutes, which is important for minimizing radiation injury to the patient's body tissues.

IA	IA	IIIB	IVB	VB	VIB	VIIB	VIIIB			IB	IIB	IIIA	IVA	VA	VIA	VIIA	VIIIA
H																	He
Li	Be											B	C	N	O	F	Ne
Na	Mg											Al	Si	P	S	Cl	Ar
K	Ca	Sc	Ti	V	Cr	Mn	Fe	Co	Ni	Cu	Zn	Ga	Ge	As	Se	Br	Kr
Rb	Sr	Y	Zr	Nb	Mo	Tc	Ru	Rh	Pd	Ag	Cd	In	Sn	Sb	Te	I	Xe
Cs	Ba	*La	Hf	Ta	W	Re	Os	Ir	Pt	Au	Hg	Tl	Pb	Bi	Po	At	Rn
Fr	Ra	†Ac	Rf	Db	Sg	Bh	Hs	Mt	Ds	Rg	Cn	Uut	Uuq	Uup	Uuh	Uus	Uuo

*	Ce	Pr	Nd	Pm	Sm	Eu	Gd	Tb	Dy	Ho	Er	Tm	Yb	Lu
†	Th	Pa	U	Np	Pu	Am	Cm	Bk	Cf	Es	Fm	Md	No	Lr

Neon, like the other noble gases in its group, is a monatomic gas that is extremely inert and unreactive. It forms no known compound with any other element. It is a colorless, odorless gas and is the fifth most common element in the atmosphere. Its concentration in air is about 0.002 percent, or approximately 0.1 that of carbon dioxide.

The familiar neon signs that we see in storefront and restaurant windows contain neon gas that glows when it is energized by an electrical discharge. When this happens, the neon atoms in the gas give off radiation in the form of orange-red light. The name "neon" sign is now often used for all such glowing tubes, although many contain gases other than neon. Different gases are used to produce signs of different colors. Every gas, when excited, radiates its own characteristic color. Argon, for example, produces a purple light, whereas xenon produces a blue-green light.

Neon was discovered in 1898 by Sir William Ramsay, an English chemist who first recognized it as an element during his experiments with the fractional distillation of liquid air. He named it after the Greek word *neos,* which means "new."

Neon signs contain neon gas that glows when energized by an electric charge.

Commercial neon is produced in large air-liquefication plants. Because neon has a boiling point of −229°C, it typically remains as a residue after the more volatile oxygen and nitrogen gases have boiled off.

Sodium

IA																	VIIIA
H	IA											IIIA	IVA	VA	VIA	VIIA	He
Li	Be											B	C	N	O	F	Ne
Na	Mg	IIIB	IVB	VB	VIB	VIIB		VIIIB		IB	IIB	Al	Si	P	S	Cl	Ar
K	Ca	Sc	Ti	V	Cr	Mn	Fe	Co	Ni	Cu	Zn	Ga	Ge	As	Se	Br	Kr
Rb	Sr	Y	Zr	Nb	Mo	Tc	Ru	Rh	Pd	Ag	Cd	In	Sn	Sb	Te	I	Xe
Cs	Ba	*La	Hf	Ta	W	Re	Os	Ir	Pt	Au	Hg	Tl	Pb	Bi	Po	At	Rn
Fr	Ra	†Ac	Rf	Db	Sg	Bh	Hs	Mt	Ds	Rg	Cn	Uut	Uuq	Uup	Uuh	Uus	Uuo

*	Ce	Pr	Nd	Pm	Sm	Eu	Gd	Tb	Dy	Ho	Er	Tm	Yb	Lu
†	Th	Pa	U	Np	Pu	Am	Cm	Bk	Cf	Es	Fm	Md	No	Lr

Atomic Number 11

Chemical Symbol Na

Group IA—The Alkali Metals

Na

Sodium is an extremely reactive, bright, silvery metal, light enough to float on water and soft enough to be cut with a knife. It is far too chemically active to be found as a pure element in nature. Even its storage presents a problem. It reacts violently with water, producing enough heat, as it sizzles and bubbles on top of the water, to melt it. It is usually kept immersed in a liquid such as kerosene to prevent it from reacting with air or moisture. Sodium is the sixth most abundant element in the Earth's crust, present at a concentration of 2.63 percent by mass.

The English chemist Sir Humphry Davy first isolated and identified sodium as an element in 1807, using an electrolytic process to separate it from molten sodium hydroxide. The name *sodium* is derived from the English word *soda*, which itself is taken from the Latin word *sodanum*, a common headache remedy of ancient times. The modern commercial method of producing sodium uses essentially the same method of electrolysis developed by Davy. The only difference is that ordinary table salt, sodium chloride, is used in place of the hydroxide. Molten sodium chloride conducts electricity, which means that the sodium and chlorine atoms must be in their electrically charged form, known as ions. As ions they can move about easily through the melt when subjected to electrical forces. When two electrodes are placed in a bath of molten sodium chloride and connected to a source of high voltage, sodium collects at one electrode, the cathode, and chlorine at the other, the anode.

Sodium has a melting point of only 98°C, which is lower than the boiling point of water. It is therefore very easy to liquefy. When sodium is transported, for example, it is pumped in liquid form into tank cars, where it solidifies. At its destination, special heating coils melt the metal and it is pumped out of the cars.

One use for liquid sodium is as a moderator in certain types of nuclear reactors. It has the right nuclear characteristics to support a chain reaction. It also has a very high specific heat, which means that a lot of heat energy is required to raise its temperature. This makes sodium an excellent material for transferring heat out of a nuclear reactor. Despite the chemical hazards of dealing with such a reactive material, reactors utilizing liquid sodium are considered safe enough to be used aboard submarines.

As a member of the alkali metals, sodium undergoes many of the chemical reactions common to these elements and is part of many important compounds that are found widely distributed throughout the Earth. The compounds of sodium are of enormous economic importance. Sodium chloride, the chemical name for common table salt, is mined in huge quantities from natural salt deposits. Seawater is another source of salt. Every gallon of seawater contains about a quarter of a pound of salt. As a nutrient and flavoring agent, table salt has been traded and bartered since early Roman times. The word *salary*, for example, is derived from the Latin word *sal* for "salt" and the fact that Roman soldiers were often given an allowance of salt as payment for their services.

Sodium hydroxide, NaOH, is one of the most important industrial compounds of sodium. It is produced commercially by the electrolysis of a solution of sodium chloride in water, also known as aqueous sodium chloride. Sodium hydroxide is a strong base that is available in most markets under the name of lye or caustic soda and serves as a drain cleaner or oven cleaner. The secret of its ability to cut grease is that it reacts with fatty material, converting this material into a substance that will dissolve in water. Soap is a mixture of fat and sodium hydroxide. In the manufacture of soap, the sodium hydroxide *saponifies* the fat, creating a water-soluble sodium "salt" of the fat. Because sodium hydroxide reacts chemically with fat, it is quite dangerous to skin and must be handled with great care.

Another important sodium compound is sodium carbonate, which is a common cleaning agent and bleach. It is known commercially as soda ash and is mixed with sand and lime to make glass. Sodium carbonate is made on a large scale by the Solvay process, in which it is manufactured from sodium chloride and limestone. The raw limestone and salt that are used in the process are inexpensive.

Another widely used compound of sodium is sodium bicarbonate, commonly known as baking soda. As its name implies,

Sodium bicarbonate, commonly known as baking soda, is used to make baked goods rise when heated or pastry dough rise when baked. It is also used as an antacid and as an agent in fire extinguishers.

Great care must be taken when storing or transporting sodium. It reacts violently when mixed with water (above). Sodium is so reactive that it is never found as a pure element in nature.

baking soda is used as a leavening agent in baking. But it also serves as an antacid to help neutralize excessive stomach acidity and as an agent in fire extinguishers, in which it is mixed with an acid to generate carbon dioxide, which smothers fires.

The isotope of sodium known as sodium-24 is radioactive and is used in biological research as a radiotracer. Because the chemistry of sodium-24 is identical to that of the stable form of sodium found in nature, the isotope can form the same compounds. Sodium chloride can be made, for example, with sodium-24 instead of ordinary sodium. In this case, scientists refer to the sodium-24 chloride molecule as being "tagged" or "labeled" with a radioactive sodium atom. When the tagged salt is ingested, its movement through the body can be traced with detectors that sense the radiation emitted by the atoms of sodium-24. Because the half-life of sodium-24 is relatively short (only 14.97 hours), the isotope dissipates quickly, minimizing the danger of radiation damage to the body.

When sodium burns, or when an electrical discharge passes through sodium vapor, it gives off a characteristic yellow light. The wavelength of this light is very precise and is used in many scientific laboratories to calibrate spectrometers and other light-measuring devices.

Sodium is also used in some types of highway lights. It is, first of all, very efficient because it gives off an intense beam of light at the expense of a relatively small amount of electricity. But highway engineers have been prompted to use sodium lights primarily because yellow light does not scatter as much in fog as other colors and because the eye is very sensitive to the color yellow. For the same reasons, many automobile manufacturers have constructed sodium fog lights for luxury cars.

IA	IA	IIIB	IVB	VB	VIB	VIIB	VIIIB			IB	IIB	IIIA	IVA	VA	VIA	VIIA	VIIIA
H																	He
Li	Be											B	C	N	O	F	Ne
Na	Mg											Al	Si	P	S	Cl	Ar
K	Ca	Sc	Ti	V	Cr	Mn	Fe	Co	Ni	Cu	Zn	Ga	Ge	As	Se	Br	Kr
Rb	Sr	Y	Zr	Nb	Mo	Tc	Ru	Rh	Pd	Ag	Cd	In	Sn	Sb	Te	I	Xe
Cs	Ba	*La	Hf	Ta	W	Re	Os	Ir	Pt	Au	Hg	Tl	Pb	Bi	Po	At	Rn
Fr	Ra	†Ac	Rf	Db	Sg	Bh	Hs	Mt	Ds	Rg	Cn	Uut	Uuq	Uup	Uuh	Uus	Uuo

*	Ce	Pr	Nd	Pm	Sm	Eu	Gd	Tb	Dy	Ho	Er	Tm	Yb	Lu
†	Th	Pa	U	Np	Pu	Am	Cm	Bk	Cf	Es	Fm	Md	No	Lr

Magnesium

Atomic Number 12

Chemical Symbol Mg

Group IIA—The Alkaline-Earth Metals

Mg

Magnesium is a silvery-white metal that is very lightweight, malleable, and moderately reactive. Its name is derived from the mineral magnesite, also known as dolomite, which consists of the white compound magnesium carbonate. Magnesite is believed to have been mined in Magnesia, a site in ancient Greece. Magnesium was discovered by the British chemist Sir Humphry Davy in 1808, using electrolysis to separate it from magnesium oxide.

Magnesium is the seventh most abundant element in the Earth's crust and is present in such large quantities in seawater that the world's oceans contain an almost unlimited supply of the dissolved metal. It is not surprising, then, that one of the most important methods of obtaining the metal, called the Dow process, involves the extraction of magnesium from seawater. In this process, oyster shells and other seashells are used to supply calcium oxide, which, when added to seawater, precipitates magnesium out of solution in the form of an insoluble solid known as milk of magnesia. After the solid is treated with hydrochloric acid to form magnesium chloride, electrolysis of the chloride yields pure magnesium metal. The molten metal is then cast into solid bars for shipment.

Magnesium has become important as a structural material. Its great advantage is that it is very light, with a density of only 1.74 grams/cm^3. For comparison's sake, water has a density of 1 gram/cm^3. Magnesium has a density that is only about one-fifth that of iron and two-thirds that of aluminum. It is usually mixed with these metals to form an alloy. When alloyed with aluminum, magnesium makes the latter metal stronger, lighter, and even more corrosion-resistant than it normally is. Many people, for example, have aluminum–magnesium alloy ladders in their homes. Its light weight also makes it ideal for fabricating automobile and aircraft parts, as well as power tools, lawn mower housings, and racing bikes.

Magnesium is the seventh most abundant element in the Earth's crust and is also present in large quantities in seawater.

Magnesium is a chemically reactive metal. In the form of a powder or metal particles, it burns brilliantly in air, giving off an intense white light that is often seen in fireworks and flares. A flashbulb consists of a thin magnesium wire that is ignited electrically by a battery. The interior of the bulb usually contains a pure oxygen atmosphere, so that the magnesium burns very rapidly. Interestingly, magnesium will also burn in carbon dioxide. Fire extinguishers that produce carbon dioxide to smother fires by preventing oxygen from getting to them would be useless in the case of a magnesium fire.

Magnesium oxide is an important compound of magnesium obtained directly from the mineral magnesite. It is used in animal feed to supply magnesium as a dietary supplement. Magnesium is also important for proper nutrition in humans because it is essential for the proper functioning of several enzymes. Additionally, it is present in chlorophyll, which plays an essential role in photosynthesis and therefore in the survival of green plants on the Earth.

Milk of magnesia is a suspension of magnesium hydroxide, $Mg(OH)_2$, in water. The familiar, creamy-looking suspension is somewhat basic and is used as an antacid because magnesium hydroxide neutralizes excess acid in the stomach.

Another common over-the-counter remedy containing magnesium is Epsom salts. These salts are a hydrated form of magnesium sulfate, which means that this molecule has several water molecules attached to it. First discovered in a well in Epsom, England, the medicinal properties of Epsom salts were noted as long ago as the early 17th century. They are still used as an aid in healing certain rashes of the skin. Magnesium is also used commercially in the tanning of leather and in the treatment of fabrics to make them accept dyes.

Relatively large concentrations of magnesium are found in some natural water supplies. This contributes to making the water "hard." The dissolved magnesium interferes with the action of detergents and forms scummy precipitates when the water is mixed with soap. The magnesium is usually removed in a process called water softening, in which it is replaced by dissolved sodium, which interferes much less with the action of detergents and soaps.

Magnesium also plays a crucial role in the makeup of the green chlorophylls present in all green plant cells. The ability of the chlorophyll to capture solar energy and convert it by photosynthesis to energy is the ultimate source of all biological energy. The structure of the chlorophyll molecule, a ring of complex atomic structures around the magnesium atom, endows the molecule with its deep color and its ability to absorb light. The role played by magnesium is to promote the formation of these light-absorbing structures.

Excess acids in the stomach can be neutralized by milk of magnesia, a suspension of magnesium hydroxide.

Aluminum

IA																	VIIIA
H	IA											IIIA	IVA	VA	VIA	VIIA	He
Li	Be											B	C	N	O	F	Ne
Na	Mg	IIIB	IVB	VB	VIB	VIIB	VIIIB			IB	IIB	Al	Si	P	S	Cl	Ar
K	Ca	Sc	Ti	V	Cr	Mn	Fe	Co	Ni	Cu	Zn	Ga	Ge	As	Se	Br	Kr
Rb	Sr	Y	Zr	Nb	Mo	Tc	Ru	Rh	Pd	Ag	Cd	In	Sn	Sb	Te	I	Xe
Cs	Ba	*La	Hf	Ta	W	Re	Os	Ir	Pt	Au	Hg	Tl	Pb	Bi	Po	At	Rn
Fr	Ra	†Ac	Rf	Db	Sg	Bh	Hs	Mt	Ds	Rg	Cn	Uut	Uuq	Uup	Uuh	Uus	Uuo

*	Ce	Pr	Nd	Pm	Sm	Eu	Gd	Tb	Dy	Ho	Er	Tm	Yb	Lu
†	Th	Pa	U	Np	Pu	Am	Cm	Bk	Cf	Es	Fm	Md	No	Lr

Atomic Number 13

Chemical Symbol Al

Group IIIA

Al

Aluminum is an extremely important commercial metal whose uses range from aluminum foil to airplane wings. As a pure element, it is a fairly soft, silvery-white metal. Usually found in nature combined with oxygen, aluminum is the third most abundant element in the Earth's crust and the most abundant metal in the crust.

Despite its abundance, aluminum remained a rather exotic and expensive metal until 1886, when Charles M. Hall in the United States and Paul L. T. Heroult in France independently discovered an inexpensive method of obtaining pure aluminum by electrolyzing aluminum oxide (Al_2O_3).

Both men discovered that aluminum oxide will dissolve in another aluminum-containing mineral called cryolite. Of vital importance is that the mixture not only conducts electricity but also has a low melting point. Pure aluminum can therefore be separated from its oxide and other compounds by passing an electric current through the molten mixture. All but a slight fraction of the world's supply of aluminum is now produced by this process. The production of aluminum consumes about 5 percent of all the electricity used in the United States today.

The principal ore of aluminum is bauxite, named after Les Baux in France, where it was discovered in 1821. This ore contains aluminum in the form of a hydrated oxide, an oxide that is combined with water molecules. Deposits of bauxite are found all over the world, but chiefly in tropical regions.

Aluminum oxide is known commercially as alumina, from which the element derives its name. The pure oxide is white and very hard. It can form crystals called corundum, one of the hardest materials known. Corundum is often used as an abrasive in

A sampling of products made from aluminum. This metal's light weight and its ability to withstand high temperatures make it an ideal ingredient for a wide range of products, from airplane wings to cans.

sandpaper and for a variety of grinding tools. It also has a very high melting point and is therefore often used to make fire bricks to line the insides of ovens and furnaces as well as the white insulating material in automobile spark plugs. In addition to these uses, it is used as a protective film on the surface of transistors and in the cosmetics industry for creams and lotions.

Crystals of corundum often contain traces of impurities that not only give it a variety of colors but also can increase its value by elevating it to the status of a gem. Some examples of gem-quality corundum are rubies (red), which contain traces of chromium, and sapphires (blue), which contain traces of iron and titanium.

Aluminum is lightweight and a good conductor of electricity, two properties that contribute to its great commercial value. Although the ability of aluminum to conduct electricity is only about 65 percent as good as that of copper, its light weight and low price make it an attractive choice for some high-voltage transmission lines. Aluminum is also an excellent reflector of radiation and is used for various types of antennas, heat reflectors, and solar mirrors.

The light weight of aluminum is a major factor in its use as a construction material. Its one drawback is that it is rather soft. This problem can be remedied by making alloys of aluminum with small quantities of copper or magnesium, which can greatly strengthen the metal.

Beyond these other properties, aluminum is also fairly reactive. When left unprotected, the surface of an aluminum object quickly becomes dull gray as the aluminum reacts with air to form an oxide. The oxide layer actually protects the metal and prevents it from further reactions with its environment, so that it is usually considered corrosion-resistant. For example, aluminum cans do not disintegrate in the way that coated steel cans do and seem ideal as beverage containers. Aluminum is also nontoxic, odorless, and tasteless, and it is a good conductor of heat, enabling beverages in aluminum containers to be cooled quickly. In a sense, it is too good a material for use in cans because its ability to resist oxidation or decay has become something of a drawback. Discarded aluminum cans that litter highways, landfills, beaches, and forests are a major environmental pollutant. Many communities have now adopted recycling as a solution to this problem.

Silicon

IA	IA	IIIB	IVB	VB	VIB	VIIB	VIIIB			IB	IIB	IIIA	IVA	VA	VIA	VIIA	VIIIA
H																	He
Li	Be											B	C	N	O	F	Ne
Na	Mg											Al	Si	P	S	Cl	Ar
K	Ca	Sc	Ti	V	Cr	Mn	Fe	Co	Ni	Cu	Zn	Ga	Ge	As	Se	Br	Kr
Rb	Sr	Y	Zr	Nb	Mo	Tc	Ru	Rh	Pd	Ag	Cd	In	Sn	Sb	Te	I	Xe
Cs	Ba	*La	Hf	Ta	W	Re	Os	Ir	Pt	Au	Hg	Tl	Pb	Bi	Po	At	Rn
Fr	Ra	†Ac	Rf	Db	Sg	Bh	Hs	Mt	Ds	Rg	Cn	Uut	Uuq	Uup	Uuh	Uus	Uuo

*	Ce	Pr	Nd	Pm	Sm	Eu	Gd	Tb	Dy	Ho	Er	Tm	Yb	Lu
†	Th	Pa	U	Np	Pu	Am	Cm	Bk	Cf	Es	Fm	Md	No	Lr

Atomic Number 14

Chemical Symbol Si

Group IVA

Si

Silicon is the second most abundant element in the Earth's crust, exceeded only by oxygen. Compounds of silicon bound chemically to oxygen make up most of the Earth's sand, rock, and soil. Silicon was first isolated in its pure form in 1824 by the Swedish chemist Jöns Jakob Berzelius (1779–1848). Today silicon forms the basis of the microelectronics industry, which for some time was centered in a region near San Francisco called Silicon Valley. The use of silicon chips in printed circuits has made possible the shrinking of room-size computers into ones that can rest on your lap.

The most important silicon compound is silica, or silicon dioxide (SiO_2). It exists in two familiar forms in nature: a crystalline form called quartz, small chips of which occur in sand, and a noncrystalline form called flint. Small gems and semiprecious stones such as amethyst, opal, agate, and jasper are crystals of quartz with colored impurities.

Flint has been known for thousands of years. When struck with an acute blow, it can be made to flake in such a manner as to produce a sharp cutting edge. Some of the earliest prehistoric tools were made of flint. Quartz crystals have a very interesting property known as the "piezoelectric effect," in which they produce an electrical current when compressed. Such crystals are used in phonograph pickups and in microphones. The opposite effect also occurs. When a quartz crystal is subjected to a vibrating electrical signal, it will exactly duplicate the vibration of the signals. Because of this effect, crystals of quartz are now so commonly used in watches and

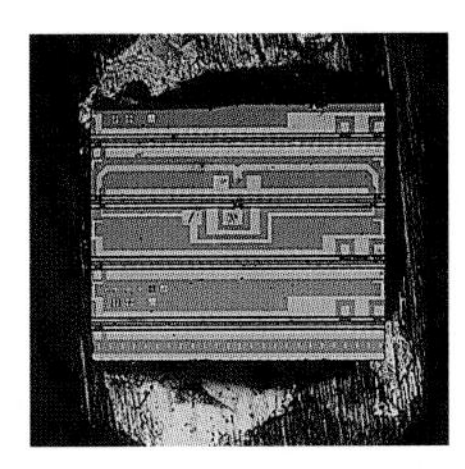

Tiny silicon chips such as this one form the basis of the microelectronics industry.

Silicon in its elementary form has the same crystal structure as diamond. This is not too surprising, because in the periodic table silicon is a member of the same group of elements as carbon.

clocks that spring-wound watches have become almost obsolete. When a quartz crystal is subjected to an alternating voltage source, it will begin to vibrate mechanically. Each quartz crystal has a natural "resonant" frequency that depends among other things on its size. One can understand resonance by thinking of a tuning fork, which only vibrates and gives off sound of a certain frequency, its resonant frequency. In a "quartz clock" this effect is used to great precision by applying electrical signals to the crystal, causing it to vibrate at its natural frequency. The vibrating crystal produces electrical signals, which are then ingeniously fed back to the crystal, keeping the frequency exactly that of the resonance. Since the frequency can be kept constant to 1 part in 10 billion, these quartz crystals are now used as time standards. Quartz crystals are also used as control devices for radio and television transmitters.

Asbestos is an important silicate mineral that forms long fibers that have great mechanical strength and resistance to heat. Until recently asbestos was used extensively as an insulating material. It is now well established that exposure to asbestos dust is very dangerous and can cause cancer of the lung and digestive tract. Many schools and factories are involved in a variety of expensive and time-consuming projects to remove asbestos from steam pipes, furnaces, automobile brakes, and a host of other products in which it was used for insulation. Prolonged exposure to silica dust itself can lead to a lung disease known as silicosis, which is a serious problem for quarry and stone workers.

Silica also produces one of the most useful and beautiful substances on Earth. When silica in the form of sand is mixed with soda ash (sodium carbonate) and limestone (calcium carbonate) and the mixture is melted, the result is glass. There are more than 1,000 different kinds of glass made for different purposes and containing various chemical compounds, but silica is the basic ingredient in most of them. The transparency of glass to light makes possible windows for our houses, eyeglasses, and optical lenses for telescopes and microscopes. Glass is also very nonreactive and resistant to attack by strong chemicals and makes an excellent material for storage jars, bottles, and food containers.

Pyrex is a type of glass made by adding an oxide of boron to the mixture of silica, soda ash, and limestone. Pyrex does not undergo as much expansion and contraction with temperature changes as does ordinary glass and is ideal for making containers that will be subjected to heat. Yet another type of glass is made by the addition of an oxide of potassium to the mixture of silica, soda ash, and limestone. This results in a particularly hard glass that is scratch resistant and is used for eyeglasses.

Ceramics are another important class of compounds based on silicon. Like glass, ceramics have been known since antiquity, during which they served as containers for food and water. Ceramics are usually made from clay. Clays are silicates that result from the weathering of granite and other rocks. Clay has a sheet-like structure that causes it to absorb large amounts of water. When fired in a furnace, it loses that water and becomes very rigid. Earthenware is very porous clay pottery that has been fired below 1,200°C. Stoneware is a ceramic that has been fired above 1,200°C. Clays are also used to make porcelain and fine china.

In 1947, scientists at Bell Laboratories invented the transistor, which incorporates layers of silicon-based semiconductors.

Silicon in its elementary form has the same crystal structure as diamond. This is not too surprising, because in the periodic table silicon is a member of the same group of elements as carbon. Pure silicon is usually prepared by heating quartz sand with carbon in the form of coke to temperatures as high as 3,000°C. Although diamond is an insulator, silicon conducts electricity to a small degree. This conducting ability can be greatly enhanced by "doping" the silicon with trace amounts of carefully chosen impurities. When a small amount of arsenic or boron is added to silicon, for example, the resultant compound is called a semiconductor. Arsenic has one more valence electron than does silicon, and when it is used to dope silicon this extra electron is free to roam about in the crystal and conduct electricity. Boron has one less electron than silicon, and this electron vacancy, or "hole," as it is called, also promotes the drift of electrons within a silicon crystal that has been doped with boron, as electrons jump from hole to hole.

Semiconductor materials made possible the invention of the transistor at Bell Laboratories in 1947. The transistor uses layers of these solid compounds to regulate the flow of electric current. Transistors have almost completely replaced vacuum tubes in modern electronic devices such as computers and television sets. Thousands of transistors can be connected together on a thin wafer of silicon.

Solar cells are also made from silicon wafers. A solar cell is composed of a silicon wafer doped with arsenic, over which is placed a thin layer of silicon doped with boron. When light falls on the wafer, electric current flows across the junction of the two materials. The silicon used in transistors and solar cells must be of extremely high purity.

Another interesting group of silicon compounds, formed by the combination of the element with organic materials, is the silicones. Silicones are polymers, or long molecular chains, made up of silicon atoms linked to oxygen atoms. Various organic compounds are often attached to the polymer to control and alter its physical properties. Silicones make up the rubbery, elastic bodies of such common toys as Silly Putty and Superball. Some silicones are designed to act as lubricants, whereas others have been used as implants in cosmetic surgery. The safety of silicone implants has been the subject of a great deal of controversy.

Phosphorus

IA	IA	IIIB	IVB	VB	VIB	VIIB	VIIIB			IB	IIB	IIIA	IVA	VA	VIA	VIIA	VIIIA
H																	He
Li	Be											B	C	N	O	F	Ne
Na	Mg											Al	Si	P	S	Cl	Ar
K	Ca	Sc	Ti	V	Cr	Mn	Fe	Co	Ni	Cu	Zn	Ga	Ge	As	Se	Br	Kr
Rb	Sr	Y	Zr	Nb	Mo	Tc	Ru	Rh	Pd	Ag	Cd	In	Sn	Sb	Te	I	Xe
Cs	Ba	*La	Hf	Ta	W	Re	Os	Ir	Pt	Au	Hg	Tl	Pb	Bi	Po	At	Rn
Fr	Ra	†Ac	Rf	Db	Sg	Bh	Hs	Mt	Ds	Rg	Cn	Uut	Uuq	Uup	Uuh	Uus	Uuo

*	Ce	Pr	Nd	Pm	Sm	Eu	Gd	Tb	Dy	Ho	Er	Tm	Yb	Lu
†	Th	Pa	U	Np	Pu	Am	Cm	Bk	Cf	Es	Fm	Md	No	Lr

Atomic Number 15

Chemical Symbol P

Group VA

Phosphorus is a nonmetal that was discovered and first isolated by the German physician Hennig Brand in 1669. Brand performed experiments in the tradition of the old alchemists. He distilled the residue from boiled-down urine and obtained something that glowed in the dark and burst into flames in warm air. What he had stumbled onto was white phosphorus, a waxy, white solid that melts at 44°C. The name given to the element was taken from the Greek word *phosphorous*, or "bearer of light."

The association of phosphorus and light is still used today when we speak of phosphorence or phosphorescent material. Technically, a material is said to be phosphorescent when it gives off light after being stimulated by radiation. The radiation can be another source of light itself. Zinc sulfide, for example, is a well-known phosphorescent material. It gives off scintillations of light when struck by fast-moving electrons. This effect on the coating of a television tube produces the TV image.

In many respects the chemistry of phosphorus resembles that of nitrogen, the element just above it in the periodic table. Like nitrogen, it is an element essential to life. About 20 percent of the human skeleton is made up of calcium phosphate. (A phosphate is a combination of phosphorus and oxygen atoms.) Our teeth are basically forms of calcium phosphate. Phosphates are also important components of DNA, the molecule that controls the genetic makeup of plants and animals.

Phosphorus occurs most commonly in nature in the form of phosphate rocks, which consist mainly of calcium phosphate and fluorapatite, a mineral containing calcium, fluorine, and phosphate groups. Phosphorus is produced commercially in quantities approaching a few million tons a year by heating calcium phosphate in the presence of carbon and silicon dioxide.

Elemental phosphorus has two important forms: white phosphorus and red phosphorus. White phosphorus is extremely reactive and spontaneously bursts into flame when exposed to air at a temperature of about 35°C. As a safety measure, it is usually stored under water. The ease with which white phosphorus burns has made it useful to the military as an incendiary device. Additional caution must be taken in handling it because it is also very toxic and can be quite damaging to human tissues, especially the cartilage and bones of the nose and jaw. At temperatures below 35°C, white phosphorus reacts with the oxygen in the air and begins to glow, a reaction that resembles phosphorescence and is known as chemiluminescence.

Soil without phosphorus is barren. Here, workers in Florida around 1890 mine phosphoric acid, which is used to make fertilizer.

Almost all of the elemental phosphorus produced commercially is used in the manufacture of phosphoric acid. Phosphoric acid is a solid that is usually sold as an 85 percent solution in water. Its major use is in the manufacture of triple-phosphate fertilizers. Soil without phosphorus is barren. Plant growth requires phosphate because so many key compounds in living plant cells are compounds of phosphorus.

In many respects the chemistry of phosphorus resembles that of nitrogen, the element just above it in the periodic table. Like nitrogen, it is an element essential to life.

The chief phosphate in detergents is sodium tripolyphosphate. It acts as a water-softening agent that inactivates magnesium and other "hard water" elements and also boosts the cleansing efficiency of detergents. Many areas ban the sale of laundry products that contain phosphate because they contribute to eutrophication, the overnourishment of plants and algae in lakes. When too many plants are allowed to grow in a relatively small body of water, the plants extract too much oxygen from the water, and fish and other marine organisms die.

Sulfur

IA	IA	IIIB	IVB	VB	VIB	VIIB	VIIIB			IB	IIB	IIIA	IVA	VA	VIA	VIIA	VIIIA
H																	He
Li	Be											B	C	N	O	F	Ne
Na	Mg											Al	Si	P	S	Cl	Ar
K	Ca	Sc	Ti	V	Cr	Mn	Fe	Co	Ni	Cu	Zn	Ga	Ge	As	Se	Br	Kr
Rb	Sr	Y	Zr	Nb	Mo	Tc	Ru	Rh	Pd	Ag	Cd	In	Sn	Sb	Te	I	Xe
Cs	Ba	*La	Hf	Ta	W	Re	Os	Ir	Pt	Au	Hg	Tl	Pb	Bi	Po	At	Rn
Fr	Ra	†Ac	Rf	Db	Sg	Bh	Hs	Mt	Ds	Rg	Cn	Uut	Uuq	Uup	Uuh	Uus	Uuo

*	Ce	Pr	Nd	Pm	Sm	Eu	Gd	Tb	Dy	Ho	Er	Tm	Yb	Lu
†	Th	Pa	U	Np	Pu	Am	Cm	Bk	Cf	Es	Fm	Md	No	Lr

Atomic Number 16

Chemical Symbol S

Group VIA

Sulfur is a reactive nonmetal found in nature both in its free elemental state and in the form of widely distributed ores and minerals. Although it is not a very abundant element, constituting only about 0.06 percent of the Earth's crust by mass, it is readily available.

In its free state, sulfur is normally a brittle, yellow solid deposited at the edges of some hot springs and geysers. It is also often found near volcanoes. Volcanic gases contain hydrogen sulfide and sulfur dioxide that react with one another to form free sulfur. It is not often recognized that volcanoes are a major source of atmospheric pollutants; however, it is estimated that volcanoes are the source of about two-thirds of the sulfur emitted into the atmosphere.

Sulfur burns with a beautiful blue flame. The old English name for sulfur was brimstone, which means "a stone that burns." The expression "fire and brimstone" is still used today to signify great heat. Solid sulfur in its crystalline form is made up of molecules that consist of eight sulfur atoms arranged in a ring.

Some common minerals of sulfur are gypsum, which is calcium sulfate, and pyrite, sometimes called "fool's gold," which is iron disulfide. Pyrite is called fool's gold because it has a golden color and often fooled novice miners into thinking they had found gold. Sulfur is also present in coal and petroleum products.

Most of the sulfur produced commercially in the United States comes from underground deposits of elemental sulfur found in Texas and Louisiana, as well as Mexico and Poland. The sulfur is mined using the Frasch process. This method was developed by the German chemical engineer Herman Frasch (1851–1914) in the 1890s. In this process pressurized water, whose temperature is above the melting point of sulfur (113°C), is pumped into a sulfur deposit to melt the element. Compressed air then forces the molten mass to the surface.

When sulfur is heated above its melting point, interesting changes occur in its molecular size. Freshly melted sulfur is a straw-colored liquid that flows very easily, almost like water. At temperatures between 160°C and 195°C, however, the viscosity of the melt increases dramatically, and it becomes approximately 100,000 times thicker. Viscosity is a measure of the ease with which liquids flow. Maple syrup flows more slowly than water, for example, and is therefore more viscous than water.

The startling change in the viscosity of sulfur is caused by the S_8 molecule changing into a molecule that may be 300,000 atoms long. These long strands get entangled with each other and resist the smooth flow of the liquid. At temperatures above 200°C, the liquid sulfur turns dark red and becomes progressively less viscous. At these high temperatures there is enough energy to break the long chains into smaller sulfur molecules, so that they no longer interfere with each other when flowing.

Because it is used in so many industrial processes, a nation's level of industrial development is often measured by its per capita consumption of sulfur.

Approximately 90 percent of all sulfur produced throughout the world is burned to form sulfur dioxide, a gas that has a characteristic suffocating odor often compared to the smell of burning matches. Most of the sulfur dioxide so produced is used to form sulfuric acid. Some, however, is used as a food preservative. Sulfur dioxide is especially toxic to molds and certain bacteria and is commonly used as an additive in wine and dried fruits. Because some people are allergic to sulfur dioxide, however, foods that contain it must be properly labeled. Sulfur dioxide is also used as a bleach for textiles and for wood pulp used in papermaking.

More sulfuric acid is produced in the United States than any other industrial chemical. A staggering 89 billion pounds of sulfuric acid were produced in 1992. This acid is the least expensive of the commercially used acids and can be prepared and shipped in pure form. Most sulfuric acid is used by the fertilizer industry to convert insoluble phosphate rock to a soluble phosphate called superphosphate, which supplies the phosphate atoms needed by growing plants. Sulfuric acid is also used for automobile batteries and in metal treatment for what is known as a "pickling process." In this process iron metal parts are "pickled" by dipping them in a sulfuric acid bath to remove rust.

Pure sulfuric acid is a colorless, oily liquid, once known as oil of vitriol. Sulfuric acid has a strong affinity for water. It removes water from many organic compounds in a reaction that often generates a great deal of heat. When the acid is poured on sugar, for example, the sugar begins to froth violently and soon yields a charred black mass of carbon. Because of its strong

Sulfur is often found deposited at the edges of some geysers, such as this one at Yellowstone National Park.

affinity for water, sulfuric acid is extremely corrosive to the skin. If you spill a drop on your skin, the chances are you will get a burn or blister no matter how quickly you flush your hand with water.

Sulfuric acid is a substantial component of the acid in acid rain. Both oil and coal contain sulfur. Coal, for example, contains pyrite, which is iron disulfide. Although the burning of oil and coal with a high sulfur content is restricted in certain areas, both oil and coal are still used as major sources of energy. When they burn, they produce sulfur dioxide, which is then further oxidized to sulfur trioxide, which in turn reacts with water to form sulfuric acid. When the sulfur trioxide reacts with water vapor, it often forms sulfuric acid aerosols, consisting of small liquid droplets of sulfuric acid held in gaseous suspension. These aerosol clouds can often persist in the atmosphere for more than a year.

More sulfuric acid is produced in the United States than any other industrial chemical.

Hydrogen sulfide is a gas that is usually said to smell like rotten eggs. The smell of rotten eggs does actually come from hydrogen sulfide, which is formed by the bacterial decomposition of sulfur-containing proteins in the yolk of the egg. Hydrogen sulfide is an extremely toxic gas, even more poisonous than the better-known cyanides. It damages the body by attacking respiratory enzymes, and even small amounts of this gas can cause headaches and nausea.

In addition to their importance in making artificial fertilizers, preserving food, bleaching textiles, and cleaning metals, sulfur compounds have hundreds of other uses in recovering metals from ores and making rubber, detergents, paints and dyes, and synthetic fibers. Indeed, a nation's level of industrial development is often measured by its per capita consumption of sulfur, and in the United States that consumption now amounts to more than 100 pounds of sulfur per person per year.

IA																	VIIIA
H	IA											IIIA	IVA	VA	VIA	VIIA	He
Li	Be											B	C	N	O	F	Ne
Na	Mg	IIIB	IVB	VB	VIB	VIIB	VIIIB			IB	IIB	Al	Si	P	S	Cl	Ar
K	Ca	Sc	Ti	V	Cr	Mn	Fe	Co	Ni	Cu	Zn	Ga	Ge	As	Se	Br	Kr
Rb	Sr	Y	Zr	Nb	Mo	Tc	Ru	Rh	Pd	Ag	Cd	In	Sn	Sb	Te	I	Xe
Cs	Ba	*La	Hf	Ta	W	Re	Os	Ir	Pt	Au	Hg	Tl	Pb	Bi	Po	At	Rn
Fr	Ra	†Ac	Rf	Db	Sg	Bh	Hs	Mt	Ds	Rg	Cn	Uut	Uuq	Uup	Uuh	Uus	Uuo

*	Ce	Pr	Nd	Pm	Sm	Eu	Gd	Tb	Dy	Ho	Er	Tm	Yb	Lu
†	Th	Pa	U	Np	Pu	Am	Cm	Bk	Cf	Es	Fm	Md	No	Lr

Chlorine

Cl

Chlorine is a poisonous, yellowish-green, diatomic gas. It was first discovered by the Swedish chemist Carl Wilhelm Scheele (1742–1786) in 1774 during his investigation of the mineral pyrolusite, but it was mistakenly identified as an oxygen-containing compound. The English chemist Sir Humphry Davy (1778–1829) finally identified chlorine as an element in 1810. Its name is derived from the Greek word *chloros*, meaning "greenish yellow."

In nature chlorine occurs mostly in dissolved salts in seawater and in the deposits in salt mines. It is an important element for the chemical industry and is among the top 10 or 15 chemicals manufactured in the United States.

Chlorine was used as a poison gas during World War I. Here, a soldier and his horse demonstrate the proper use of gas masks.

During World War I, chlorine was used as a poison gas on the battlefields of Europe. Chlorine is extremely dangerous, and inhaling even a small amount can cause extensive lung damage. The toxicity of chlorine makes it an excellent disinfectant for swimming pools and water supplies.

The bleaching action of chlorine is one of its best known properties. The development of an effective bleach has a long history. Early attempts at dissolving chlorine in water were not very successful. The solution formed an acid called hydrochloric acid, which unfortunately had

Atomic Number **17**

Chemical Symbol **Cl**

Group VIIA—The Halogens

Chlorine was first discovered by the Swedish chemist Carl Wilhelm Scheele in 1774, but he mistakenly identified it as an oxygen-containing compound. The English chemist Sir Humphry Davy finally identified chlorine as an element in 1810.

a tendency to dissolve cotton and linen. After many years of research the familiar liquid bleach used in households today was developed. This bleach effectively contains a weak solution of sodium hypochlorite. It is the hypochlorite (OCl) that does the bleaching.

The large amounts of chlorine needed for industry are produced by the electrolysis of sodium chloride dissolved in water. A large amount of the chlorine commercially produced is used for the manufacture of polymers such as polyvinyl chloride. Polyvinyl chloride, or PVC as it is called, is a plastic that has replaced iron as a favored material for constructing waste and water pipes because it does not oxidize or easily corrode. It is also used to manufacture the colorless plastic bottles that have replaced glass as containers for most soft drinks.

An important compound of chlorine is hydrogen chloride (HCl). It is a colorless gas with a sharp, penetrating odor. About 3 million tons are produced annually. Hydrogen chloride gas dissolves very easily in water to produce a solution known as hydrochloric acid. This acid is used as a solvent and for removing rust from steel a process called "pickling." This is usually done before galvanizing steel or electroplating it with zinc to resist corrosion. Hydrochloric acid is also the acid present in the gastric juices of the stomach, where it is needed to activate protein-digesting enzymes.

Some well-known commercial solvents that are compounds of chlorine include carbon tetrachloride and chloroform. Chloroform is a volatile liquid that was long used as an anesthetic for surgery. Because of evidence that it can severely damage the liver and kidneys, it has been replaced by other compounds.

Carbon tetrachloride, or CCl_4, is an excellent solvent for dissolving greases and oils. Until quite recently it was extensively used in the dry cleaning of clothes. Studies have shown, however, that carbon tetrachloride is extremely toxic to the liver, and its use is now illegal in many areas.

Large amounts of chlorine have also been used to produce insecticides such as DDT (dichlorodiphenyltrichloroethane). Many of these compounds have, however, been banned because they are considered harmful to the environment. Other chlorine compounds known to cause environmental problems are the chlorofluorocarbons. As almost ideal refrigerants, they were until recently universally used in all types of air conditioners and refrigerators. Studies have shown, however, that chlorofluorocarbons generate chlorine "free radicals" in the upper layers of the atmosphere, which destroy the ozone layer that protects humans from harmful solar radiation. The use of chlorofluorocarbons is now strictly controlled by an international agreement known as the Montreal Protocol.

IA	IA	IIIB	IVB	VB	VIB	VIIB	VIIIB			IB	IIB	IIIA	IVA	VA	VIA	VIIA	VIIIA
H																	He
Li	Be											B	C	N	O	F	Ne
Na	Mg											Al	Si	P	S	Cl	Ar
K	Ca	Sc	Ti	V	Cr	Mn	Fe	Co	Ni	Cu	Zn	Ga	Ge	As	Se	Br	Kr
Rb	Sr	Y	Zr	Nb	Mo	Tc	Ru	Rh	Pd	Ag	Cd	In	Sn	Sb	Te	I	Xe
Cs	Ba	*La	Hf	Ta	W	Re	Os	Ir	Pt	Au	Hg	Tl	Pb	Bi	Po	At	Rn
Fr	Ra	†Ac	Rf	Db	Sg	Bh	Hs	Mt	Ds	Rg	Cn	Uut	Uuq	Uup	Uuh	Uus	Uuo

*	Ce	Pr	Nd	Pm	Sm	Eu	Gd	Tb	Dy	Ho	Er	Tm	Yb	Lu
†	Th	Pa	U	Np	Pu	Am	Cm	Bk	Cf	Es	Fm	Md	No	Lr

Atomic Number **18**

Chemical Symbol **Ar**

Group **VIIIA—The Noble Gases**

Ar

Argon is a noble gas that is chemically inert. It has no color and is odorless. It constitutes about 1 percent of the Earth's atmosphere, making it the third most abundant gas in the air. It was long thought to form no compounds, but a group of Finnish researchers demonstrated in August 2000 that despite its lack of chemical reactivity it was nevertheless possible to combine argon with other atoms to form a compound. This new compound is extremely fragile and can exist only at the extremely low temperature of –265°C. It is formed with fluorine and hydrogen and known as argon fluorohydride, or HArF.

Argon was the first noble gas to be discovered. It was identified in 1894 by the English physicist Lord Rayleigh and the Scottish chemist William Ramsay. They identified the new element by means of the light it gave off when an electrical discharge was passed through the gas. They called it argon, from the Greek word *argos*, meaning "lazy" or "inactive."

In 1894, English physicist Lord Rayleigh (above), working with William Ramsay, discovered argon, the first noble gas to be identified.

The discovery of argon was a beautiful example of scientific detective work. Lord Rayleigh was working with two samples of nitrogen. One was obtained from the air and the other from decomposing a nitrogen-containing compound such as ammonia. He noticed that there was a slight difference in the density of the two samples—a difference of only 0.05

Incandescent light bulbs are filled with argon instead of ordinary air because argon is a noble gas and, unlike oxygen, will not react with the tungsten filament.

percent. Rather than dismiss the results as an experimental error, he looked for the cause of the difference.

The essential clue to this mystery was that the air sample had been obtained by removing all the gases known at that time to exist in air, such as oxygen, carbon dioxide, and water vapor. To account for the difference in density, Rayleigh, with the help of Ramsay, correctly guessed that the air sample must contain a small amount of an unknown gas heavier than nitrogen. Since it had escaped detection for so long, he also realized that it must be a chemically unreactive gas. This new gas was argon.

A new column in the periodic table, Group VIIIA, was created to place argon among the other elements. The elements in this column are now known as the noble gases. Lord Rayleigh was awarded the 1904 Nobel Prize in physics for his discovery of argon.

Commercially, argon is obtained by the distillation of liquid air. The boiling point of argon is between that of nitrogen and oxygen, and argon is usually a by-product of the production of these two other gases.

The commercial applications of argon make use of its lack of chemical reactivity. It is a relatively abundant gas, so that it is more commonly used than many of the other noble gases. Argon is used to fill incandescent light bulbs, for example, replacing ordinary air. This prevents corrosion of the tungsten filament in these bulbs and inhibits its vaporization, which can cause blackening of the bulbs. Argon is also used to supply an inert environment during welding. Fluorescent bulbs also contain argon instead of air.

Argon has also been used in laboratories engaged in nuclear physics. The probe that senses the presence of radiation in a Geiger counter, for example, is often filled with argon. Radiation passing through the window of the probe ionizes the argon, causing the probe to emit a small electrical discharge that can be detected.

Argon is also the decay product of an important radioisotope used for dating rock samples. The isotope is potassium-40, a naturally occurring radioactive isotope with an extremely long half-life of 1.25 billion years. Argon is formed as the product of the disintegration of the radioactive potassium. The technique is called potassium-argon dating. Because potassium is present in many rocks, an analysis of the amount of argon present in the rock will effectively reveal the date at which the rock solidified.

IA	IA	IIIB	IVB	VB	VIB	VIIB	VIIIB			IB	IIB	IIIA	IVA	VA	VIA	VIIA	VIIIA
H																	He
Li	Be											B	C	N	O	F	Ne
Na	Mg											Al	Si	P	S	Cl	Ar
K	Ca	Sc	Ti	V	Cr	Mn	Fe	Co	Ni	Cu	Zn	Ga	Ge	As	Se	Br	Kr
Rb	Sr	Y	Zr	Nb	Mo	Tc	Ru	Rh	Pd	Ag	Cd	In	Sn	Sb	Te	I	Xe
Cs	Ba	*La	Hf	Ta	W	Re	Os	Ir	Pt	Au	Hg	Tl	Pb	Bi	Po	At	Rn
Fr	Ra	†Ac	Rf	Db	Sg	Bh	Hs	Mt	Ds	Rg	Cn	Uut	Uuq	Uup	Uuh	Uus	Uuo

*	Ce	Pr	Nd	Pm	Sm	Eu	Gd	Tb	Dy	Ho	Er	Tm	Yb	Lu
†	Th	Pa	U	Np	Pu	Am	Cm	Bk	Cf	Es	Fm	Md	No	Lr

K

Potassium is a silvery-white metal with a putty- or wax-like consistency that is so soft it can easily be cut with a knife. It is a member of the alkali metals, the group that contains such elements as lithium and sodium. Like them, it is extremely reactive and so is never found in the free state in nature.

Potassium occupies the position just below sodium in the periodic table. It is therefore not surprising that sodium and potassium are chemically rather similar and about equally abundant in nature. Both of these elements occur in silicate minerals and in seawater. Interestingly, the oceans contain much more sodium than potassium. Potassium is essential for plant growth, whereas sodium is not, so that plants take up much of the potassium in minerals as it filters, dissolved in water, through soils before entering streams and rivers on its way to the sea.

Potassium was first isolated by Sir Humphry Davy in 1807. As in so many of his previous discoveries, Davy used electrolysis as a means of separating elements from their compounds.

Almost all of the potassium chloride that is mined is used as plant fertilizer. In fact, plants and trees themselves were an early source of potassium for human use. Wood and other plant materials were burned in pots to give an ash, called potash (potassium-rich ash), which consists primarily of potassium carbonate. The name *potassium* has its origin in the word *potash* from this early source of the element.

Like all of the alkali metals, potassium reacts violently with water to produce hydrogen. This reaction generates so much heat that it can ignite the hydrogen that bubbles off, causing flames to be produced. To prevent this from happening, it is

Modern high-powered rifle bullets no longer use black gunpowder made with potassium, which produces a great deal of smoke, leaves a heavy residue, and requires frequent cleaning of the rifle bore. Gun cartridges use "smokeless" powder today, which consists basically of nitrocellulose.

usually stored immersed in a liquid such as kerosene or naphtha.

Potassium burns in air to produce potassium superoxide, whose chemical formula is KO_2. This is an interesting compound that reacts with both water and carbon dioxide to produce oxygen. These properties of potassium superoxide are utilized in self-contained breathing devices. They permit a diver to breathe naturally, using the oxygen generated internally by the superoxide from exhaled carbon dioxide, without any exposure to outside fumes.

Several compounds of potassium are of commercial interest. Potassium hydroxide, KOH, is an extremely strong base that is very soluble in water. It is used chiefly as an electrolyte in certain types of storage batteries and in the manufacture of liquid soap.

Potassium nitrate, KNO_3, is an important compound that has been known for centuries. It is better known as saltpeter, which really means "rocksalt." The name derives from the Greek word *petra*, for rock. It resembles ordinary table salt in appearance. When saltpeter is dissolved in water, it has a slightly salty taste, which explains its name. It is used as a preservative and as an important component of potassium-containing fertilizers. Perhaps its most spectacular use is as an explosive. Potassium nitrate decomposes when heated, releasing large quantities of nitrogen gas. Gunpowder consists of potassium nitrate, wood charcoal, and sulfur. When gunpowder is heated, large volumes of carbon dioxide and nitrogen gas are released. The sudden expansion of these hot gases causes an explosion.

A naturally occurring isotope of potassium is the radioactive isotope potassium-40. It occurs naturally in many rocks and has an unusually long half-life of 1.25 billion years. Potassium-40 is used extensively to date rocks. This technique depends on the fact that when potassium-40 decays, it transforms itself into the noble gas argon. Consequently, to determine the age of a rock, one need only determine how much argon is present in the rock. The oldest rocks on Earth have been dated by this method as being 3.8 billion years old.

Potassium-40 is also an important source of normal background radiation. Each human body contains about 140 grams of potassium distributed throughout the body. Since the natural abundance of potassium-40 is about 0.012 percent, we are all partially made up of this radioactive isotope. There is no escaping its radiation, and it is a major contributor to our lifetime dose of radiation.

IA	IA	IIIB	IVB	VB	VIB	VIIB	VIIIB			IB	IIB	IIIA	IVA	VA	VIA	VIIA	VIIIA
H																	He
Li	Be											B	C	N	O	F	Ne
Na	Mg											Al	Si	P	S	Cl	Ar
K	Ca	Sc	Ti	V	Cr	Mn	Fe	Co	Ni	Cu	Zn	Ga	Ge	As	Se	Br	Kr
Rb	Sr	Y	Zr	Nb	Mo	Tc	Ru	Rh	Pd	Ag	Cd	In	Sn	Sb	Te	I	Xe
Cs	Ba	*La	Hf	Ta	W	Re	Os	Ir	Pt	Au	Hg	Tl	Pb	Bi	Po	At	Rn
Fr	Ra	†Ac	Rf	Db	Sg	Bh	Hs	Mt	Ds	Rg	Cn	Uut	Uuq	Uup	Uuh	Uus	Uuo

*	Ce	Pr	Nd	Pm	Sm	Eu	Gd	Tb	Dy	Ho	Er	Tm	Yb	Lu
†	Th	Pa	U	Np	Pu	Am	Cm	Bk	Cf	Es	Fm	Md	No	Lr

Calcium

Atomic Number 20

Chemical Symbol Ca

Group IIA—The Alkaline-Earth Metals

Ca

Calcium is the fifth most abundant element in the Earth's crust, and it is an especially important nutrient for a wide range of living organisms. Human teeth and bones contain calcium, and marine organisms build their shells of calcium carbonate ($CaCO_3$), which also makes up the hardened portion of the coral reefs of the Bahamas and the Florida Keys. When these coral organisms die, the sediments of their shells form the large deposits of limestone found distributed throughout the Earth. Limestone is chiefly calcium carbonate and is a rich source of marine fossils.

Calcium is far too active an element to be found as a pure metal in nature. It reacts with moisture to form calcium hydroxide ($Ca(OH)_2$) and with the oxygen in air to form calcium oxide (CaO). Pure calcium metal is fairly hard and has the characteristic silvery-white color of all the alkaline-earth metals.

As far back as the 1st century, the Romans used a compound of calcium known today as lime. They called it *calx*, the Latin name for lime, from which the modern name for calcium is derived. The metal was first isolated and identified as an element by Sir Humphry Davy in 1808. Pure calcium metal has rather limited commercial use and is produced in small quantities by heating calcium oxide with aluminum.

Many water supplies contain dissolved calcium in the form of calcium ions (Ca^{2+}). The calcium ion has lost two of the valence electrons of the calcium atom and so is positively charged. In this form it can easily react with other chemical groups. The calcium ion is known as one of the "hardness ions" because its presence produces hard water. Water is said to be hard when an ion such as calcium combines with soap to form

Calcium is a fairly reactive metal. Here it is shown reacting with water to produce hydrogen gas.

an insoluble scumlike precipitate. The result is to decrease the ability of soap to remove dirt and grease.

Another problem arises in regions where hard water also contains bicarbonate ions $(HCO_3)^-$. When such water is heated, the dissolved bicarbonate ions undergo a chemical reaction that forms carbon dioxide gas (CO_2) and carbonate ions $(CO_3)^=$. Because gases become less soluble in water as the temperature is raised, the heat drives the gaseous carbon dioxide out of solution, causing more bicarbonate to break down into carbon dioxide and carbonate. In the presence of calcium ions, insoluble calcium carbonate ($CaCO_3$) is formed in the water. This insoluble substance, or precipitate, sticks to the walls of boilers and hot water pipes, narrowing them and restricting the flow of water. In boilers, the precipitate is known as "boiler scale" and is a very serious problem because the scale interferes with the conduction of heat through the walls of the boiler.

Some compounds of calcium are better known by their popular names than by their chemical formulas. Limestone, for example, is calcium carbonate; quicklime, or simply lime, is calcium oxide; and slaked lime is calcium hydroxide. Many compounds of calcium are extremely important commercially. Calcium carbonate is used as an antacid. Mortar is made by mixing slaked lime with sand and water. As the mixture dries, the calcium hydroxide crystallizes and the mortar hardens. Eventually, the mortar reacts with the carbon dioxide in the air to form an extremely hard matrix of limestone and sand. It is used for bricklaying. The ancient Romans used mortar to construct buildings and roads.

Another ancient building material is concrete. It has its origins in Egypt and has been in constant use during the past 5,000 years. Most concretes used today are based on Portland cement. Discovered by an English bricklayer in 1824, Portland cement owes its name to its resemblance to the natural limestone found on the Isle of Portland in England. The cement is made from a mixture of limestone, sand, clay, and gypsum. When it is mixed with sand and water, it hardens into the familiar material we see in so many of the structures of our daily life.

Lime (calcium oxide) is an extremely important industrial chemical. It is easily made by heating limestone. One of the early uses of lime was in theatrical lighting. When lime is heated to a high temperature it gives off an intense, bluish-white light. Such light was used in the early 19th century to illuminate actors, giving rise to the phrase "in the limelight."

Probably the most important modern use of lime is in the production of iron from its ores. The process of making iron

A limestone cave at Carlsbad Caverns National Park in New Mexico.

usually begins with the addition of a mixture of iron ore and lime to a blast furnace. As the mixture is heated, the lime combines with impurities in the iron ore to form a glassy material called slag. The molten slag then flows to the bottom of the furnace and is separately removed as the molten iron pours from the furnace.

Limestone caves are among the most impressive naturally occurring structures. These caves are formed slowly over thousands of years when slightly acidic groundwater, formed by the presence of carbon dioxide in the water, seeps through cracks in the rocks and dissolves enough limestone to hollow out a large opening. The limestone often reprecipitates to form the icicle-like formations known as stalactites and stalagmites. Stalactites grow downward from a cave's ceiling and stalagmites grow upward from a cave's floor (an easy way to remember this is to associate the *c* in stalactite with "ceiling" and the *g* in stalagmite with "ground").

Marble is composed of calcium carbonate and is very sensitive to acid rain. Many marble structures, such as statues, columns, and the facades of public buildings, have been badly damaged by acid rain. Governments dedicated to preserving structures that are of historic interest or that have some artistic merit have been working very hard to control the emissions that produce acid rain.

Gypsum, the popular name for calcium sulfate dihydrate ($CaSO_4 \cdot 2H_2O$), is an important mineral derived from the sea. The word *dihydrate* refers here to the two water molecules attached to every molecule of calcium sulfate in gypsum. A

Calcium is an especially important nutrient for a wide range of living organisms. Human teeth and bones contain calcium, and marine organisms build their shells of calcium carbonate.

familiar white chalky material, gypsum was formed when the great inland seas and lakes that once dotted the Earth dried up eons ago. It is widely distributed in nature, constituting most of the White Sands National Monument in New Mexico, for example. It is also a highly important building material and is used to make a variety of products that everyone has seen or used. It forms the plaster used to coat the walls and ceilings of houses as well as the plaster casts used to set broken bones and the plaster molds for artists' sculptures.

The versatility of gypsum is based on the loss of some of its water molecules when it is heated, giving rise to a new material called plaster of paris. When water is added to the powdered plaster of paris, it recreates small crystals of gypsum, which join with one another to form a hard mass of gypsum. This reaction happens very rapidly and produces a great deal of heat.

Gypsum also exists in the crystalline form known as alabaster. This material is very soft and easy to carve, making it a favorite material of many sculptors. When polished, alabaster becomes translucent, which adds greatly to its beauty.

Calcium chloride ($CaCl_2$) is a compound of calcium that has a strong affinity for water and can actually absorb water from the air, often absorbing enough to dissolve itself. Compounds of this kind are called deliquescent. Many commercial products contain calcium chloride because of its ability to remove moisture from damp places such as basements.

IA	IA	IIIB	IVB	VB	VIB	VIIB	VIIIB			IB	IIB	IIIA	IVA	VA	VIA	VIIA	VIIIA
H																	He
Li	Be											B	C	N	O	F	Ne
Na	Mg											Al	Si	P	S	Cl	Ar
K	Ca	Sc	Ti	V	Cr	Mn	Fe	Co	Ni	Cu	Zn	Ga	Ge	As	Se	Br	Kr
Rb	Sr	Y	Zr	Nb	Mo	Tc	Ru	Rh	Pd	Ag	Cd	In	Sn	Sb	Te	I	Xe
Cs	Ba	*La	Hf	Ta	W	Re	Os	Ir	Pt	Au	Hg	Tl	Pb	Bi	Po	At	Rn
Fr	Ra	†Ac	Rf	Db	Sg	Bh	Hs	Mt	Ds	Rg	Cn	Uut	Uuq	Uup	Uuh	Uus	Uuo

*	Ce	Pr	Nd	Pm	Sm	Eu	Gd	Tb	Dy	Ho	Er	Tm	Yb	Lu
†	Th	Pa	U	Np	Pu	Am	Cm	Bk	Cf	Es	Fm	Md	No	Lr

Scandium

Atomic Number 21

Chemical Symbol Sc

Group IIIB—First-Row Transition Element

Sc

Scandium heads a list of 10 metals called the first-row transition elements, which occupy the center of the periodic table. As the number of protons in the transition elements increases across the row of these elements, electrons are added to an incompletely filled inner shell rather than to the outer, valence shell. Consequently, the number of valence electrons is virtually the same for all of these elements. This similarity in valence electron configuration causes the transition elements to resemble each other in their chemical behavior. All of them, for example, are fairly unreactive metals, and many of their compounds are colored. Care must be taken in dealing with these elements and their compounds because many of them are extremely hazardous. Elements such as chromium, nickel, cobalt, zinc, and titanium, for example, are known to cause cancer.

Scandium is a scarce element that makes up approximately 0.0025 percent of the Earth's crust. Interestingly, it is found in greater concentrations in the sun and certain other stars. It was one of the elements that Mendeleyev, in 1871, predicted theoretically would fill one of the vacant gaps in his periodic table. The element was discovered 8 years later, in 1879, by Lars Fredrik Nilson, who named it after his homeland, Scandinavia. The discovery of scandium led to widespread acceptance of the predictive power of the periodic table.

Scandium is a very lightweight metal with a fairly high melting point and good resistance to corrosion. These properties have made it of great interest to the aerospace industry for the construction of aircraft. The pure metal was not available until 1937, and some of the first samples of the silvery-white metal

The discovery of scandium in 1879 led to widespread acceptance of the predictive power of the periodic table.

were produced for the U.S. Air Force. Pure scandium is usually prepared by the electrolysis of scandium chloride ($ScCl_2$). It has a tendency to develop a yellowish color when exposed to air.

Scandium forms very few useful compounds. The metal itself has found some use in electronic devices, such as high-intensity lamps that produce light with a color value close to that of natural sunlight. Lamps of this kind are often used to illuminate baseball and football stadiums.

One of the isotopes of scandium, known as scandium-46, has found some use as a tracer in the refining of petroleum. Scandium-46 is radioactive, with a half-life of 83.8 days. When added during the crude-oil refining process, it can keep track of, or trace, by means of its radioactivity, certain desirable components. This greatly increases the efficiency of separating the crude oil into useful components.

IA	IA	IIIB	IVB	VB	VIB	VIIB	VIIIB			IB	IIB	IIIA	IVA	VA	VIA	VIIA	VIIIA
H																	He
Li	Be											B	C	N	O	F	Ne
Na	Mg											Al	Si	P	S	Cl	Ar
K	Ca	Sc	Ti	V	Cr	Mn	Fe	Co	Ni	Cu	Zn	Ga	Ge	As	Se	Br	Kr
Rb	Sr	Y	Zr	Nb	Mo	Tc	Ru	Rh	Pd	Ag	Cd	In	Sn	Sb	Te	I	Xe
Cs	Ba	*La	Hf	Ta	W	Re	Os	Ir	Pt	Au	Hg	Tl	Pb	Bi	Po	At	Rn
Fr	Ra	†Ac	Rf	Db	Sg	Bh	Hs	Mt	Ds	Rg	Cn	Uut	Uuq	Uup	Uuh	Uus	Uuo

*	Ce	Pr	Nd	Pm	Sm	Eu	Gd	Tb	Dy	Ho	Er	Tm	Yb	Lu
†	Th	Pa	U	Np	Pu	Am	Cm	Bk	Cf	Es	Fm	Md	No	Lr

Titanium

Atomic Number 22

Chemical Symbol Ti

Group IVB—First-Row Transition Element

Ti

Titanium is the ninth most abundant element in the Earth's crust, making up some 0.6 percent of its mass. In its pure state it is a shiny, white metal that is easy to work and quite ductile, or capable of being drawn into wire. Discovered in 1791 by an amateur English mineralogist, the Reverend William Gregor, titanium is named for the Titans of Greek mythology. The minerals rutile and ilemnite are its usual sources. The most common technique at the present time of winning the metal from its ore is chemical reduction. This is a difficult and expensive process because titanium reacts readily with oxygen and carbon at high temperatures. In September 2000, however, a group of metallurgists, headed by George Chen of Cambridge University, reported an ingenious new way to extract titanium from its ore by electrolysis. If this technique can be scaled up to produce industrial amounts, it would greatly reduce the cost of the metal.

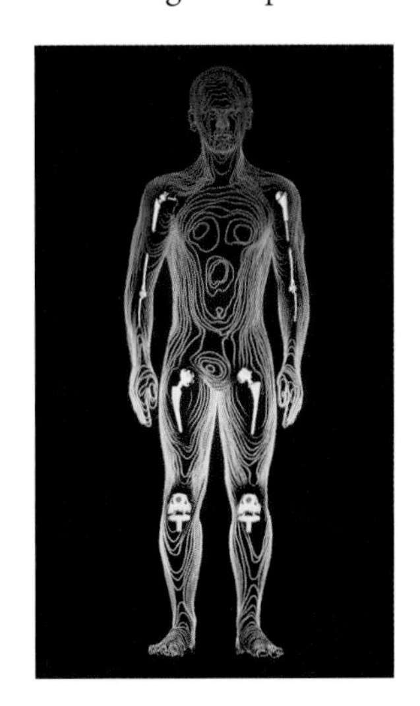

Because titanium does not react with human tissue, titanium pins are often used in surgery.

Titanium has a density approximately 40 percent that of steel. Despite its light weight, it is unusually strong and virtually immune to the usual kinds of metal fatigue. It also has an extraordinary resistance to corrosion, so that it has every property needed to make it an ideal structural material for rockets and jet

Titanium's strength and light weight make it an ideal component of aircraft engines.

engines. To take but one example, some 10,000 pounds of titanium and its compounds are used in each engine of a Boeing 747 jet.

Titanium bicycles have recently taken the sporting world by storm. Many cyclists claim that their combination of light weight and stiffness produces an almost effortless ride. These state-of-the-art bikes are, unfortunately, very expensive. Titanium's resistance to chemical attack has also made it a valuable tool in medicine. Titanium pins, for example, are often used in bone surgery because they do not react with tissue.

The most important compound of titanium is titanium dioxide (TiO_2), a substance with a brilliant, intense white color that is used as a pigment for paints, paper, and plastics. It has also found use as a sunscreen because it can prevent damage to the skin without the use of any other chemical. The annual production of titanium dioxide in the United States is approximately 1 million tons.

Another useful compound of titanium is titanium tetrachloride ($TiCl_4$), a colorless liquid. When exposed to moist air, it forms a dense white cloud of small titanium dioxide particles. The U.S. Navy used titanium tetrachloride during World War II to make smoke screens when it was necessary to block a potential target from view.

IA																	VIIIA
H	IA											IIIA	IVA	VA	VIA	VIIA	He
Li	Be											B	C	N	O	F	Ne
Na	Mg	IIIB	IVB	VB	VIB	VIIB		VIIIB		IB	IIB	Al	Si	P	S	Cl	Ar
K	Ca	Sc	Ti	V	Cr	Mn	Fe	Co	Ni	Cu	Zn	Ga	Ge	As	Se	Br	Kr
Rb	Sr	Y	Zr	Nb	Mo	Tc	Ru	Rh	Pd	Ag	Cd	In	Sn	Sb	Te	I	Xe
Cs	Ba	*La	Hf	Ta	W	Re	Os	Ir	Pt	Au	Hg	Tl	Pb	Bi	Po	At	Rn
Fr	Ra	†Ac	Rf	Db	Sg	Bh	Hs	Mt	Ds	Rg	Cn	Uut	Uuq	Uup	Uuh	Uus	Uuo

*	Ce	Pr	Nd	Pm	Sm	Eu	Gd	Tb	Dy	Ho	Er	Tm	Yb	Lu
†	Th	Pa	U	Np	Pu	Am	Cm	Bk	Cf	Es	Fm	Md	No	Lr

Vanadium

Atomic Number **23**

Chemical Symbol **V**

Group VB—First-Row Transition Element

V

Vanadium is a bright, shiny metal that is fairly soft and extremely resistant to corrosion. It makes up about 0.02 percent of the Earth's crust, and it is found in trace quantities in more than 60 different minerals. As a source of the metal, the most important of these minerals is vanadinite.

A Mexican professor of mineralogy, Andres Manuel del Rio, discovered vanadium in 1801. It was later named for the Scandinavian goddess Vanadis because of its many beautifully colored compounds.

About 80 percent of the vanadium produced in the United States is used as an additive in the making of steel. Its general effect is to make the steel more resistant to wear and stress and also to make the steel perform better at high temperatures. Fortunately, this application does not require very pure vanadium, which is difficult to prepare in quantity because of its reactivity with oxygen and carbon at high temperatures. For this reason, a compound of vanadium known as ferrovanadium, rather than the pure metal, is added to iron to form vanadium steel, a hard steel used for engine parts and cutting tools.

The most important compound of vanadium is its oxide. This yellowish-red crystal is known chemically as vanadium pentoxide (V_2O_5). It is used commercially as a catalyst in the contact process for preparing sulfuric acid. In the contact process, large amounts of sulfur trioxide (SO_3), an important intermediate compound in the manufacture of sulfuric acid, are produced. The sulfur trioxide is usually made by exposing sulfur dioxide (SO_2) to oxygen. But under ordinary conditions, this reaction proceeds very slowly. If these materials are combined

A Mexican professor of mineralogy, Andres Manuel del Rio, discovered vanadium in 1801.

while in contact with solid vanadium oxide, however, the sulfur trioxide is produced very rapidly. The vanadium oxide acts as a catalyst; it does not enter into the chemical reaction but simply speeds it up.

Vanadium is often used as an alloying material in the steels that make up the structural parts of nuclear reactors. Ordinary steel has a tendency to "creep," that is, become distorted and stretched, when it is subjected to high temperatures or weights over a long period of time. In a reactor, for example, this could cause the rods containing the fuel elements to rupture. Vanadium steels are much more resistant to creep. Another advantage is that vanadium does not readily absorb neutrons. Neutrons are the nuclear particles that fission the uranium in a reactor and perpetuate the chain reaction. The presence of vanadium, then, has very little effect on the nuclear processes occurring in the reactor.

IA	IA	IIIB	IVB	VB	VIB	VIIB	VIIIB			IB	IIB	IIIA	IVA	VA	VIA	VIIA	VIIIA
H																	He
Li	Be											B	C	N	O	F	Ne
Na	Mg											Al	Si	P	S	Cl	Ar
K	Ca	Sc	Ti	V	Cr	Mn	Fe	Co	Ni	Cu	Zn	Ga	Ge	As	Se	Br	Kr
Rb	Sr	Y	Zr	Nb	Mo	Tc	Ru	Rh	Pd	Ag	Cd	In	Sn	Sb	Te	I	Xe
Cs	Ba	*La	Hf	Ta	W	Re	Os	Ir	Pt	Au	Hg	Tl	Pb	Bi	Po	At	Rn
Fr	Ra	†Ac	Rf	Db	Sg	Bh	Hs	Mt	Ds	Rg	Cn	Uut	Uuq	Uup	Uuh	Uus	Uuo

*	Ce	Pr	Nd	Pm	Sm	Eu	Gd	Tb	Dy	Ho	Er	Tm	Yb	Lu
†	Th	Pa	U	Np	Pu	Am	Cm	Bk	Cf	Es	Fm	Md	No	Lr

Chromium was discovered by French chemist Louis-Nicolas Vauquelin in 1797 and named from the Greek word *chroma*, meaning "color." This was an appropriate choice because all of the compounds of chromium are colored. The pure metal, however, has a silvery-white color. Although brittle, it is quite hard.

Chromium metal is usually extracted from chromite, an oxide of chromium that is its most important ore. Chromium oxide (Cr_2O_3) is among the 10 most abundant compounds in the Earth's crust. When exposed to air, chromium forms an invisible oxide that makes the metal extremely resistant to corrosion and very useful as both a decorative and a protective coating over other metals such as brass, bronze, and steel. Almost everyone is familiar with the chrome plate that is deposited electrostatically on automobile parts such as bumpers. Chromium is also used in large amounts to produce stainless steel, a steel alloy that is resistant to corrosion. A typical stainless steel object can contain as much as 18 percent chromium.

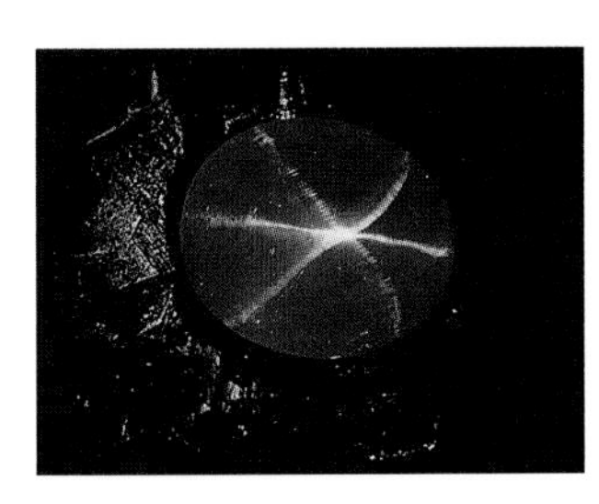

The brilliant red color of rubies can be traced to the presence of small amounts of chromium in the mineral corundum.

One of the principal uses of chromium compounds is in the manufacture of pigments. Chromium oxide (Cr_2O_3), for example, is the most stable green pigment known and is used for coloring paints, cements, and plaster. Another chromium-containing pigment is the yellow pigment known as lead chromate yellow. Pigments account for about 35 percent of the chromium-containing chemicals produced each year.

The beautiful color of many precious gems is also caused by the presence of trace amounts of chromium. The red color of rubies, for example, is caused by small amounts of chromium present in the mineral corundum. The characteristic green color of emeralds is caused by trace amounts of chromium in the mineral beryl. One of the most interesting precious stones is alexandrite, which contains a small amount of chromium in its basic mineral, chrysoberyl. Alexandrite has the property of changing color in different light sources. When viewed in the light coming from a fire, for example, it is a deep red color, whereas in daylight it is a beautiful blue.

Chromium was discovered by a French chemist in 1797 and named from the Greek word chroma, *meaning "color."*

In addition to their use for treating the surfaces of metals so as to render them resistant to corrosion, chromium compounds are used for tanning leathers and for producing high-quality recording tapes. Electrical engineers have found that the electrical and magnetic properties of chromium oxide produce a superior sound, and it is the active material coated onto these tapes.

In 1960, the first laser, a ruby laser, was constructed by the American physicist Theodore H. Maiman. The ruby laser consists of a ruby crystal, essentially aluminum oxide (Al_2O_3) that contains a small amount of chromium atoms as an impurity in the crystal. The action of a laser is described in the section on helium. In the ruby laser, it is the chromium that stores the energy and is responsible for the laser action. The ruby crystal is usually in the shape of a cylindrical rod with partially mirrored ends, surrounded by a tube containing xenon gas. When the xenon tube flashes and gives off light, the light stimulates the chromium atoms to release their energy. The mirrors reflect some of the radiation back into the tube, which has the effect of further stimulating the release of energy by the chromium atoms.

Chromium compounds should be handled with care because many of them are toxic.

IA	IA	IIIB	IVB	VB	VIB	VIIB	VIIIB			IB	IIB	IIIA	IVA	VA	VIA	VIIA	VIIIA
H																	He
Li	Be											B	C	N	O	F	Ne
Na	Mg											Al	Si	P	S	Cl	Ar
K	Ca	Sc	Ti	V	Cr	Mn	Fe	Co	Ni	Cu	Zn	Ga	Ge	As	Se	Br	Kr
Rb	Sr	Y	Zr	Nb	Mo	Tc	Ru	Rh	Pd	Ag	Cd	In	Sn	Sb	Te	I	Xe
Cs	Ba	*La	Hf	Ta	W	Re	Os	Ir	Pt	Au	Hg	Tl	Pb	Bi	Po	At	Rn
Fr	Ra	†Ac	Rf	Db	Sg	Bh	Hs	Mt	Ds	Rg	Cn	Uut	Uuq	Uup	Uuh	Uus	Uuo

*	Ce	Pr	Nd	Pm	Sm	Eu	Gd	Tb	Dy	Ho	Er	Tm	Yb	Lu
†	Th	Pa	U	Np	Pu	Am	Cm	Bk	Cf	Es	Fm	Md	No	Lr

Manganese

Atomic Number **25**

Chemical Symbol **Mn**

Group VIIB—First-Row Transition Element

Mn

Manganese is a hard, gray-white metal that has many properties similar to those of iron, its neighbor in the periodic table. It not only looks like iron, but also like iron corrodes in moist air. Manganese metal does not exist in its free state in nature and is most commonly found combined with oxygen in pyrolusite, a mineral that consists chiefly of manganese dioxide (MnO_2). Manganese was discovered by Carl Wilhelm Scheele in 1774 and named from the Latin word *magnes,* meaning "magnet," referring to the fact that pyrolusite has magnetic properties.

The United States has virtually no sources of manganese, but rich ores containing as much as 40 percent manganese are mined in such areas as South Africa and India. An unusual source of manganese is the so-called manganese nodules scattered on the bottom of the sea. These nodules, roughly spherical in shape and containing chiefly manganese and iron oxides, are thought to have been produced by microorganisms that have the ability to extract manganese from seawater.

When manganese is added to steel, it forms an unusually hard steel that is very resistant to shock. This makes such steel ideal for use in rifle barrels, bank vaults, railroad tracks, and earth-moving equipment. Manganese also adds hardness, strength, and corrosion resistance to alloys of aluminum and magnesium.

One of the most common compounds of manganese in chemical laboratories around the world is potassium permanganate ($KMnO_4$), which is used as an indicator for acid solutions. When dissolved in water, it forms a deep-purple solution. If an acid is then added to the solution, its color quickly turns from purple to a very pale pink. This ability to change color is thus used to "indicate" a change to an acidic state. Potassium permanganate is also used to purify public water supplies and to absorb toxic gases.

The purplish color of permanganate is sometimes seen in antique glass. In the early 19th century, glassmakers added manganese oxides to their glass to eliminate the yellow tinge produced by iron impurities that were present. With age and long exposure to the sun, the manganese became converted into a permanganate, giving the glass a faint purplish gleam. This coloring, present in old windows, is now highly prized by owners of early American homes throughout New England. Although glass manufacturers no longer add manganese, its ability to color objects is still used to brighten ceramics and pottery.

Manganese was discovered by Carl Wilhelm Scheele in 1774 and named from the Latin word magnes, *meaning "magnet."*

Manganese dioxide is a compound commonly used in the dry-cell batteries found in most inexpensive flashlights. The technical name for these batteries is zinc–carbon dry cells. The anode, or positive terminal of the battery, which is the little button on top, is connected to a carbon electrode surrounded by a moist paste that contains manganese, among other compounds. The complex chemical reactions that occur in this paste are essential for the operation of the battery.

The manganese essentially removes hydrogen gas that forms around the carbon electrode as the battery is being discharged. The hydrogen is a product of the chemical reactions that produce the electricity. This production of hydrogen gas is known as polarization, and if the hydrogen were not removed, the battery would soon cease to function. Since a gas is a poor conductor of electricity, the hydrogen buildup is disruptive because it prevents the proper transfer of electrons to the electrode. The common alkaline battery also uses manganese dioxide as one of its reactants.

Iron

IA																	VIIIA
H	IA											IIIA	IVA	VA	VIA	VIIA	He
Li	Be											B	C	N	O	F	Ne
Na	Mg	IIIB	IVB	VB	VIB	VIIB	VIIIB			IB	IIB	Al	Si	P	S	Cl	Ar
K	Ca	Sc	Ti	V	Cr	Mn	Fe	Co	Ni	Cu	Zn	Ga	Ge	As	Se	Br	Kr
Rb	Sr	Y	Zr	Nb	Mo	Tc	Ru	Rh	Pd	Ag	Cd	In	Sn	Sb	Te	I	Xe
Cs	Ba	*La	Hf	Ta	W	Re	Os	Ir	Pt	Au	Hg	Tl	Pb	Bi	Po	At	Rn
Fr	Ra	†Ac	Rf	Db	Sg	Bh	Hs	Mt	Ds	Rg	Cn	Uut	Uuq	Uup	Uuh	Uus	Uuo

*	Ce	Pr	Nd	Pm	Sm	Eu	Gd	Tb	Dy	Ho	Er	Tm	Yb	Lu
†	Th	Pa	U	Np	Pu	Am	Cm	Bk	Cf	Es	Fm	Md	No	Lr

Atomic Number 26

Chemical Symbol Fe

Group VIIIB—First-Row Transition Element

Fe

Iron is probably the most common metal in human society. It plays a major role in almost every aspect of everyday life. Whether we are using a screwdriver or a washing machine or riding in an automobile or a train, the importance and usefulness of iron as a structural material are apparent.

Iron is the fourth most abundant element in the Earth's crust and, next to aluminum, the second most abundant metal. The interior of the Earth, known as the core, is thought to be made chiefly of molten iron.

Iron is never found in its pure state in nature. Most iron-bearing ores contain iron combined with oxygen in the form of oxides. Two of the most important iron-containing minerals are hematite and magnetite.

In its pure state, iron is a somewhat lustrous, silvery-white metal. Unfortunately, the pure metal quickly corrodes. It reacts with water and air to form rust. Rust is a hydrated oxide of iron ($Fe_2O_3 \cdot xH_2O$), which means that the iron oxide molecule has some water molecules attached to it (the *x* in the formula indicates a variable or unknown number of molecules of water hydration). Rust is a porous, reddish material that adheres very poorly to the surface of iron and usually crumbles off. Unfortunately, the ease with which it does this constantly exposes fresh iron to oxidation, so that many iron artifacts quickly disintegrate if not protected.

Iron was discovered thousands of years ago. The ability to refine the metal from the reddish ores in which it is found served as a major milestone in human development, known as the Iron Age (about 1100 B.C.). The discovery of iron led to tools

This blast furnace in Sparrows Point, Maryland, is capable of making 8,000 tons of iron per day.

and weapons that were harder and more durable than those known during the previous Bronze Age (about 3000 B.C.). Today more than 90 percent of all metal refined in the world is iron.

The production of iron usually takes place in a blast furnace. There, iron ore and carbon in the form of coke are heated to convert iron oxide in the ore into molten iron. This molten iron, known as "pig iron," contains many impurities such as carbon, manganese, and silicon, and it is usually too brittle for most applications. Removing most of the carbon by further heating with large blasts of oxygen converts the pig iron to the much harder form of iron known as steel.

There are many kinds of steel. Each kind is made by adding a small amount of some specific impurity that gives the steel a desired property. The steel is usually named for this impurity. Nickel steel, for example, which contains a small amount of added nickel, is very resistant to tensile, or pulling, stresses. It is used for building bridges, power-transmission towers, and ordinary bicycle chains. Tungsten and vanadium steels are particularly useful for tools that must maintain their cutting edges at high temperatures. Manganese steel can withstand high-energy impacts very well and is therefore used to manufacture rifle barrels and power shovels for moving earth. Stainless steels are made by adding chromium and nickel to iron, often in concentrations as high as 12 to 18 percent.

Iron plays a crucial part in the transport of oxygen throughout the body. The role played by hemoglobin, the molecule in red blood cells that carries oxygen from the lungs to all the tissues of the body, has been discussed in the section on oxygen. It is the four iron atoms that each hemoglobin molecule contains, however, that are central to its structure and function. The oxygen molecules actually bond to the iron atoms and are then transported by the blood to cells in need of oxygen. It is worth noting that carbon monoxide, a common gas produced in combustion, is bound 200 times more strongly than oxygen to the iron atoms. This means that in an environment of high carbon monoxide concentration, asphyxiation (carbon monoxide poisoning) can occur, since the hemoglobin will become saturated with carbon monoxide and little hemoglobin will be available for oxygen transport.

Today more than 90 percent of all metal refined in the world is iron.

An important property associated with iron is its magnetism. Iron is attracted to magnets, and magnetite, a compound of iron, is itself strongly magnetic. This property is caused by the structure of the iron atom. The electrons orbiting the nucleus of the iron atom make each such atom a tiny magnet.

The fact that an electric current will produce a magnetic field is well known. It is this effect, for example, that is at the heart of an electric motor. But an electric current is really the drift of electrons through a wire. The electrons orbiting the nucleus of an iron atom resemble an electric current, and this motion therefore makes each atom a tiny magnet. In addition, the electrons are spinning on their own axes, very much like the way the Earth spins on its axis. This motion also enhances the magnetic properties of the atom.

In the metal itself, the spacing of the atoms is just right to cause their individual magnetic forces to enhance each other and produce a strong magnetic effect. Heating or melting can destroy this spacing and cause the loss of magnetism.

Besides its presence on Earth, iron is the heaviest element that can be made in the interior of stars through the ordinary process of fusion. In a star, hydrogen nuclei can be compressed together to form all of the elements up to and including iron, but the formation of iron absorbs more energy than it releases, and the fusion process stops. Elements more complex than iron are formed during the extraordinary explosions of dying stars known as supernovas.

Cobalt

IA	IA	IIIB	IVB	VB	VIB	VIIB	VIIIB			IB	IIB	IIIA	IVA	VA	VIA	VIIA	VIIIA
H																	He
Li	Be											B	C	N	O	F	Ne
Na	Mg											Al	Si	P	S	Cl	Ar
K	Ca	Sc	Ti	V	Cr	Mn	Fe	Co	Ni	Cu	Zn	Ga	Ge	As	Se	Br	Kr
Rb	Sr	Y	Zr	Nb	Mo	Tc	Ru	Rh	Pd	Ag	Cd	In	Sn	Sb	Te	I	Xe
Cs	Ba	*La	Hf	Ta	W	Re	Os	Ir	Pt	Au	Hg	Tl	Pb	Bi	Po	At	Rn
Fr	Ra	†Ac	Rf	Db	Sg	Bh	Hs	Mt	Ds	Rg	Cn	Uut	Uuq	Uup	Uuh	Uus	Uuo

*	Ce	Pr	Nd	Pm	Sm	Eu	Gd	Tb	Dy	Ho	Er	Tm	Yb	Lu
†	Th	Pa	U	Np	Pu	Am	Cm	Bk	Cf	Es	Fm	Md	No	Lr

Atomic Number 27

Chemical Symbol Co

Group VIIIB—First-Row Transition Element

Co

Cobalt is a rather rare element, making up only 0.003 percent of the Earth's crust. One of its major ores is called cobaltite, a compound of cobalt, arsenic, and sulfur. The pure metal, with its characteristic bright bluish-white color, is obtained by roasting this ore in air. Cobalt metal was first isolated by the Swedish chemist Georg Brandt in 1739.

The fact that glass made with trace amounts of cobalt minerals is blue was known to the ancient Egyptians thousands of years ago. The name *cobalt* derives from the German word *kobold*, which refers to an evil gnome or spirit. Miners often said problems or accidents occurring in the mines were caused by "kobald." The mining of cobalt ores was apparently no exception, and the name *cobalt* became the accepted one for the mineral in the 18th century.

Cobalt is often added to steel to improve its resistance to corrosion. When cobalt is mixed with tungsten and copper, it forms an alloy called stellite, a metal that retains its hardness even at high temperatures. This makes it ideal for high-speed drills and cutting instruments.

Like iron, elemental cobalt is known for its magnetic properties. It is easily magnetized and retains its magnetism at temperatures far higher than those for iron. Some 25 percent of all the cobalt produced in the world is used to make a powerfully magnetic substance called alnico. Alnico is an alloy of aluminum, nickel, and cobalt and is used for making industrial magnets. Its name consists of the first two letters of each of its component elements.

When naturally occurring cobalt is placed in a nuclear reactor and exposed to large numbers of neutrons, it can be transformed

When cobalt is mixed with tungsten and copper, it forms an alloy called stellite, a metal that remains hard even at high temperatures, making it ideal for use in high-speed cutting instruments.

into a highly radioactive isotope known as cobalt-60. This isotope emits powerful gamma rays that are similar to X rays. The gamma rays are used for treating certain types of cancer and for sterilizing food.

Cobalt-60 has a half-life of 5.2 years, so that once formed it will be radioactive for a relatively long time. Unlike a massive and complicated X-ray machine with its cumbersome power source, the isotope, when properly shielded, is easy to transport from one location to another. This makes it particularly useful for field applications, such as using its gamma rays to look for cracks in the hulls of ships.

Cobalt is known to be one of the trace elements essential to human nutrition. It is present in meat and dairy products and in vitamin B-12, an important substance that prevents the disease known as pernicious anemia, in which the blood is depleted of adequate numbers of its oxygen-carrying red cells.

Cobalt is known to be one of the trace elements essential to human nutrition. It is present in meat and dairy products and in vitamin B-12.

Nickel

IA																	VIIIA
H	IA											IIIA	IVA	VA	VIA	VIIA	He
Li	Be											B	C	N	O	F	Ne
Na	Mg	IIIB	IVB	VB	VIB	VIIB		VIIIB		IB	IIB	Al	Si	P	S	Cl	Ar
K	Ca	Sc	Ti	V	Cr	Mn	Fe	Co	Ni	Cu	Zn	Ga	Ge	As	Se	Br	Kr
Rb	Sr	Y	Zr	Nb	Mo	Tc	Ru	Rh	Pd	Ag	Cd	In	Sn	Sb	Te	I	Xe
Cs	Ba	*La	Hf	Ta	W	Re	Os	Ir	Pt	Au	Hg	Tl	Pb	Bi	Po	At	Rn
Fr	Ra	†Ac	Rf	Db	Sg	Bh	Hs	Mt	Ds	Rg	Cn	Uut	Uuq	Uup	Uuh	Uus	Uuo

*	Ce	Pr	Nd	Pm	Sm	Eu	Gd	Tb	Dy	Ho	Er	Tm	Yb	Lu
†	Th	Pa	U	Np	Pu	Am	Cm	Bk	Cf	Es	Fm	Md	No	Lr

Ni

Nickel is a silvery metal found chiefly in the ore called millerite, a compound of nickel and sulfur. It is considered a rare element because only 0.01 of 1 percent of the earth's crust is nickel. Although nickel is scarce on the surface of the Earth, many scientists believe that large deposits of this metal exist deep within its interior. In fact, the molten core of the Earth is thought to be composed chiefly of iron and nickel. This may explain why nickel is often found in meteorites, since these extraterrestrial rock fragments are thought to have been formed at about the same time as the Earth.

Meteorites, such as this one found in Arizona, often contain nickel.

Atomic Number **28**

Chemical Symbol **Ni**

Group **VIIIB—First-Row Transition Element**

Although nickel compounds have been known since ancient times, the pure metal was first isolated by a Swedish chemist named Axel Fredrik Cronstedt in 1751. *Nickel* is a German word for Satan, and the element is thought to have been named for the German *Kupfernickel*, or Satan's copper. Like cobalt, nickel compounds were known to the ancient world as a way of coloring glass. The characteristic color that nickel compounds add to glass and other substances is green.

Nickel is extremely resistant to corrosion and is frequently added to other metals to form alloys resistant to oxidation. Nickel plating, often called electroplating, is a technique that adds a protective coating of nickel to the surface of a metal, such as iron or steel, that is known to corrode fairly easily.

Stainless steel, which typically contains about 18 percent chromium and 8 percent nickel, provides another example of the use of nickel to prevent corrosion. An alloy known as monel is a mixture of nickel and copper whose hardness and resistance to corrosion make it the metal of choice for such applications as the propeller shafts of boats.

Nichrome, the familiar metal used to make the heating elements in toasters and electric ovens, is an alloy of chromium and nickel. The high electrical resistance of nichrome, combined with its high melting point, make it a very efficient material for converting electrical energy into heat energy.

Perhaps the most obvious use of nickel is in the U.S. five-cent coin, which is called the nickel. The coin is actually made of copper and nickel, with nickel making up approximately 25 percent of the alloy. Like iron and cobalt, nickel can be made magnetic. The alloy named alnico, formed with aluminum, nickel, and cobalt, is used to create some of the most powerful magnets known.

An important modern use of nickel is in the nickel–cadmium battery. One of the electrodes in this battery is a form of nickel oxide. This battery is rechargeable, which makes it particularly useful in calculators, computers, and cordless electric shavers, for example. The nickel–cadmium battery can also be sealed to prevent leakage, which is of great importance in electronic equipment. It produces a power output of 1.4 volts or only slightly less than the 1.5 volts produced by an ordinary dry cell.

The U.S. five-cent coin, called the nickel, is actually made primarily of copper. Only 25 percent of the alloy is nickel.

Nickel *is a German word for Satan, and the element is thought to have been named for the German* **Kupfernickel,** *or Satan's copper.*

Copper

IA	IA	IIIB	IVB	VB	VIB	VIIB	VIIIB			IB	IIB	IIIA	IVA	VA	VIA	VIIA	VIIIA
H																	He
Li	Be											B	C	N	O	F	Ne
Na	Mg											Al	Si	P	S	Cl	Ar
K	Ca	Sc	Ti	V	Cr	Mn	Fe	Co	Ni	Cu	Zn	Ga	Ge	As	Se	Br	Kr
Rb	Sr	Y	Zr	Nb	Mo	Tc	Ru	Rh	Pd	Ag	Cd	In	Sn	Sb	Te	I	Xe
Cs	Ba	*La	Hf	Ta	W	Re	Os	Ir	Pt	Au	Hg	Tl	Pb	Bi	Po	At	Rn
Fr	Ra	†Ac	Rf	Db	Sg	Bh	Hs	Mt	Ds	Rg	Cn	Uut	Uuq	Uup	Uuh	Uus	Uuo

*	Ce	Pr	Nd	Pm	Sm	Eu	Gd	Tb	Dy	Ho	Er	Tm	Yb	Lu
†	Th	Pa	U	Np	Pu	Am	Cm	Bk	Cf	Es	Fm	Md	No	Lr

Atomic Number **29**

Chemical symbol **Cu**

Group **IB—First-Row Transition Element (Coinage Metal)**

Cu

Copper is one of the most familiar metals of everyday life. Look at the plumbing in your home and you are likely to see copper pipes carrying water to your kitchen and bathroom. Because copper is one of the best conductors of electricity, copper wires are widely used to transmit electrical energy from power stations to homes, offices, factories, and other buildings and from wall outlets to electrical appliances. The concentration

Brass, an alloy of copper and zinc, has a wide variety of uses, from hardware to ammunition.

of electrical wires under a major city such as New York rivals the mass of copper in many mines.

Its mechanical properties also make copper almost ideal for electrical transmission. A soft metal, it is extremely ductile, which means that it has the ability to be drawn into wire. Commercial copper wire is made by drawing the metal through a series of dies of decreasing diameter so that the wire becomes thinner with each die.

Another common object, the "copper" penny used in the United States, is unfortunately no longer made of pure copper. Since 1981, the penny has been treated with copper plating to give the coin its characteristic reddish-brown color. Copper is used as a coinage metal because it is relatively unreactive with air and water. It was also once used to make buttons for uniform jackets worn by policemen, and the common slang expression "copper" refers to this practice.

Copper was once used to make buttons for uniform jackets worn by policemen. The common slang expression "copper" refers to this practice.

We may commonly think of corrosion as a process producing rust on objects made of iron. However, copper, although relatively unreactive, also corrodes slowly in the atmosphere by forming a layer either of copper carbonate ($CuCO_3$) or copper sulfate ($CuSO_4$). These green substances are said to form a "patina" that protects the copper metal underneath them from further corrosion. A dramatic example of the formation of patina is the Statue of Liberty, made in France of copper plates. Before being cleaned for its centennial, it was almost entirely green. Many people consider the patina that covers old church steeples or ancient monuments quite beautiful and have resisted attempts to remove it. When modern architects choose copper plates for the roofs and facades of buildings, it is with the expectation that the copper will corrode and produce a beautiful green color.

Despite the common use of copper, it is a rather rare element, making up only 0.007 percent of the Earth's crust. When found in its pure metallic state in nature, it is often called "native" copper. More common sources of copper are ores such as chalcopyrite, in which the element is combined with sulfur and iron. Most copper is currently obtained by the open-pit mining of low-grade ores containing only a small percentage of copper. The pure metal is obtained either by roasting the ore, in which it is heated in air, or by electrolysis.

Copper derives its name from the Latin word *cuprum*, which means "from Cyprus," an obvious reference to the island that served as a source of the metal for the Romans. It is easy to refine copper from its ore, and it was therefore known to many ancient civilizations. Copper jewelry dating back to 9000 B.C. has been found in Iraq. Because it is a relatively soft metal, copper never found widespread use for weapons or tools.

When copper is combined with other metals to form alloys, the result is often a new metal that is harder and tougher than any of the original metals used in the alloy. Brass is essentially an alloy of copper and zinc, whereas bronze is an alloy of copper and tin. Bronze tools and weapons served human beings before they learned to smelt iron. When nickel is combined with copper, it produces a hard and strong alloy called monel that resists corrosion.

Copper in large amounts is quite toxic, and copper compounds are used to kill various forms of fungi and bacteria. Paints used for ships' hulls, for example, contain copper to prevent the excessive growth of marine organisms on the surface of the hull. This growth is called fouling and develops slowly on any surface exposed to seawater.

Copper's name comes from the Latin word cuprum, *which means "from Cyprus," a reference to the island that served as a source of the metal for the Romans.*

The greenish tint on the Statue of Liberty is a result of the corrosion of its copper plates.

IA	IA	IIIB	IVB	VB	VIB	VIIB	VIIIB			IB	IIB	IIIA	IVA	VA	VIA	VIIA	VIIIA
H																	He
Li	Be											B	C	N	O	F	Ne
Na	Mg											Al	Si	P	S	Cl	Ar
K	Ca	Sc	Ti	V	Cr	Mn	Fe	Co	Ni	Cu	Zn	Ga	Ge	As	Se	Br	Kr
Rb	Sr	Y	Zr	Nb	Mo	Tc	Ru	Rh	Pd	Ag	Cd	In	Sn	Sb	Te	I	Xe
Cs	Ba	*La	Hf	Ta	W	Re	Os	Ir	Pt	Au	Hg	Tl	Pb	Bi	Po	At	Rn
Fr	Ra	†Ac	Rf	Db	Sg	Bh	Hs	Mt	Ds	Rg	Cn	Uut	Uuq	Uup	Uuh	Uus	Uuo

*	Ce	Pr	Nd	Pm	Sm	Eu	Gd	Tb	Dy	Ho	Er	Tm	Yb	Lu
†	Th	Pa	U	Np	Pu	Am	Cm	Bk	Cf	Es	Fm	Md	No	Lr

Zinc

Zn

Zinc is not a very abundant element, making up about 0.007 percent of the Earth's crust. It is found chiefly in the mineral zinc sulfide, also known as sphalerite or zincblende. Commercial use of zinc dates back to the 15th century, but there is no record of when the metal was discovered.

Although zinc is fairly reactive in its pure state, in which it is a hard, brittle, silvery metal, it is relatively resistant to corrosion and quickly forms a hard oxide coating that prevents it from reacting further with the air. In the process known as galvanization, a layer of zinc is coated over steel to protect it from corrosion. Galvanized steel is commonly used for a variety of household objects such as metal garbage pails and chain-link fences. The zinc is applied either by dipping a steel object into some molten zinc or using an electrolytic process to deposit the zinc on the steel. The zinc coating prevents air and moisture from coming into contact with the steel; even if the zinc coating is scratched or broken, the steel is still protected by a phenomenon called galvanic protection, in which the zinc acts as a sacrificial metal. This is because zinc is more easily oxidized than iron, and if corrosion occurs, the zinc rather than the iron reacts. If a zinc disc is attached to the iron rudder of a ship, for example, the zinc will gradually corrode and disappear, but the iron rudder will remain unattacked.

Although the U.S. penny is coated with a thin sheet of copper to retain its original color, since 1981 it has been made primarily of zinc.

Atomic Number 30

Chemical Symbol Zn

Group IIB—First-Row Transition Element

Although some 90 percent of the zinc produced in the United States is used for galvanizing steel, the metal has many other uses. One of the most important is in the common dry-cell battery. This source of energy for flashlights, portable radios, and radio-controlled toys was invented more than 100 years ago by the French chemist George Leclanche. The battery consists of a zinc inner case, just beneath an outer protective metal shell that serves as one electrode (the anode) and a carbon rod that serves as the other electrode (the cathode). It generates an electrical force of about 1.5 volts.

Since 1981, zinc has also served as the chief metal in the U.S. penny, although this coin is still covered with a thin sheet of copper to retain its original "copper" color. Zinc is also combined with copper to form the hard and durable alloy called brass.

Some other useful zinc compounds include zinc oxide, a white powder that is made by burning zinc vapor in air. One of the many uses of zinc oxide is as a pigment in making white paints. In the form of an ointment, zinc oxide has become a popular sunscreen, blocking the sun's ultraviolet light and preventing it from damaging the skin. Zinc oxide is also photoconductive. This means that it conducts electricity when exposed to light. In a photocopier, a zinc oxide plate is electrically charged and exposed to a printed document. When a light is then passed through the document and onto the plate, the parts of the plate that are lighted will carry the electrical charge away from the plate, whereas the dark parts, corresponding to the ink on the document, remain charged. When a black powder called a toner is then distributed over the surface, it sticks to the electrically charged parts of the plate, reproducing the image of the document. This image is then transferred to the paper by heating.

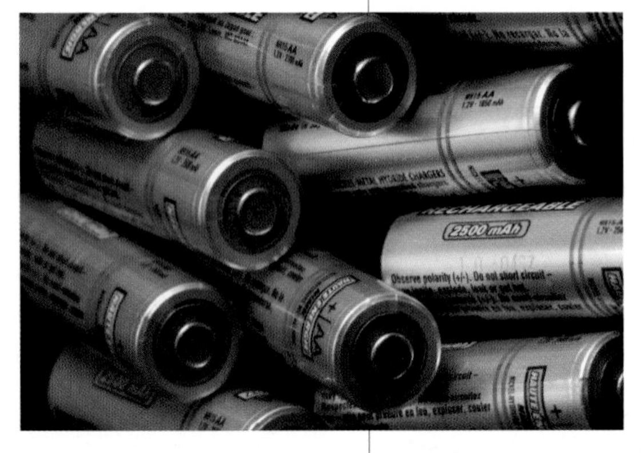

The common dry-cell battery used to power flashlights, toys, and radios usually consists of a zinc outer cylinder (the anode) and an inner carbon rod (the cathode).

Another compound of zinc, zinc sulfide, is used in many electronic devices as a phosphor. Phosphors have the important property of giving off light when struck by electrons. Among other uses, zinc sulfide is used to coat the inner surfaces of television tubes and of the cathode-ray tubes used as monitors in computers. When a beam of electrons generated in the tube strikes the sulfide coating, it produces the picture or text you see on these screens.

IA																	VIIIA
H	IA											IIIA	IVA	VA	VIA	VIIA	He
Li	Be											B	C	N	O	F	Ne
Na	Mg	IIIB	IVB	VB	VIB	VIIB	VIIIB			IB	IIB	Al	Si	P	S	Cl	Ar
K	Ca	Sc	Ti	V	Cr	Mn	Fe	Co	Ni	Cu	Zn	Ga	Ge	As	Se	Br	Kr
Rb	Sr	Y	Zr	Nb	Mo	Tc	Ru	Rh	Pd	Ag	Cd	In	Sn	Sb	Te	I	Xe
Cs	Ba	*La	Hf	Ta	W	Re	Os	Ir	Pt	Au	Hg	Tl	Pb	Bi	Po	At	Rn
Fr	Ra	†Ac	Rf	Db	Sg	Bh	Hs	Mt	Ds	Rg	Cn	Uut	Uuq	Uup	Uuh	Uus	Uuo

*	Ce	Pr	Nd	Pm	Sm	Eu	Gd	Tb	Dy	Ho	Er	Tm	Yb	Lu
†	Th	Pa	U	Np	Pu	Am	Cm	Bk	Cf	Es	Fm	Md	No	Lr

Gallium

Atomic Number 31

Chemical Symbol Ga

Group IIIA—Post-Transition Metal

Ga

Gallium belongs to the same group of elements as aluminum and shares many chemical properties with that element. It is an extremely soft metal that can be cut with a knife. It also has an extremely low melting point of 29.8°C. Since the temperature of the human body is 37°C, gallium will melt when held in the palm of the hand. However, since it also has the extremely high boiling point of 2,403°C, the range of temperatures within which gallium is liquid is the largest of any known metal. This makes gallium useful for special high-temperature thermometers. Like water, it also has the unusual ability to expand when it freezes.

Commercially, gallium is ordinarily generated as a by-product in the refining of other metals. Chief among these is aluminum, with which it is present in the mineral bauxite. The history of gallium is of interest because it is one of the elements that Mendeleyev predicted theoretically in 1871. Using this prediction, the French chemist Paul-Émile Lecoq de Boisbaudran found and identified the element in 1875. He named it for his native France, using the Latin name, Gallia, for that country.

Light-emitting diodes (LEDs), made of gallium arsenide, are found in a wide range of electronic products.

Until quite recently gallium was a rather exotic element with few practical applications. This changed very rapidly with the discovery that gallium arsenide (GaAs), a compound of gallium and arsenic,

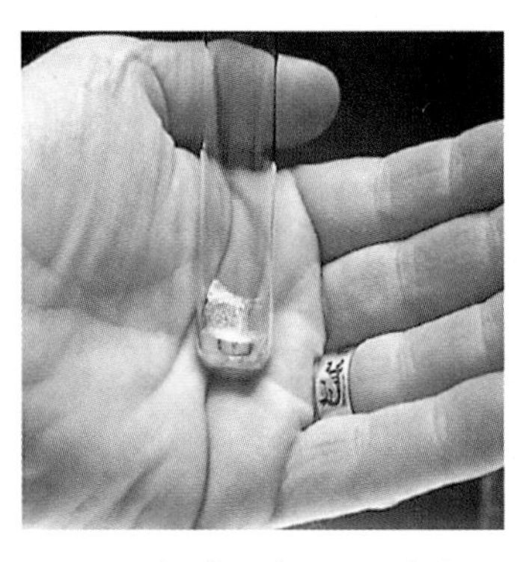

Because of gallium's extremely low melting point, when a test tube filled with gallium is placed in the palm of a person's hand, the metal will start to melt.

could function as a "laser diode" and convert electricity directly into a beam of laser light. It is used in preparing lasers for compact disc players. Light-emitting diodes, or LEDs, made of gallium arsenide are used in a variety of electronic displays, watches, and autodisc players.

Besides converting electricity to light, gallium arsenide is a semiconductor. It generates less heat than the customary silicon transistor chips used in computers, and GaAs chips are replacing silicon in many supercomputers that require higher power.

Gallium-67 was one of the first artificially produced radioactive isotopes to be used in medicine. Because it has a half-life of only 78 hours, it rapidly loses its radioactivity and ability to harm the body. The isotope has a tendency to concentrate in the tissues of certain cancers, such as melanoma, and can therefore be used to treat the disease without doing too much damage to surrounding tissue.

A French chemist found and identified gallium in 1875. He named it for his native country using the Latin name for France, Gallia.

Gallium is also at the heart of two of the world's major neutrino detectors engaged in efforts to detect neutrinos reaching Earth from the Sun. Neutrinos are subatomic particles that are extremely elusive and difficult to detect. They can pass through miles of solid rock without interacting. The usefulness of gallium is based on the fact that the collision of a neutrino with a gallium nucleus can often convert the gallium into the radioactive isotope germanium-71, which is easy to detect and identify.

One of the detectors is buried deep underground in the Gran Sasso tunnel in Italy. The other, known as SAGE, is also underground and is found in the small town of Baksan in the Caucasus Mountains in Russia. SAGE is an acronym for Soviet–American Gallium Experiment and contains approximately 250,000 pounds of gallium, which at its present selling price of $390 a pound represents a major resource for the current Russian government. Its value has threatened to undermine the research efforts of the facility. Not only has the government threatened to seize the gallium, but also there have been several attempts by armed bands of thieves to steal the valuable metal from the underground laboratory.

Germanium

IA	IA	IIIB	IVB	VB	VIB	VIIB	VIIIB			IB	IIB	IIIA	IVA	VA	VIA	VIIA	VIIIA
H																	He
Li	Be											B	C	N	O	F	Ne
Na	Mg											Al	Si	P	S	Cl	Ar
K	Ca	Sc	Ti	V	Cr	Mn	Fe	Co	Ni	Cu	Zn	Ga	Ge	As	Se	Br	Kr
Rb	Sr	Y	Zr	Nb	Mo	Tc	Ru	Rh	Pd	Ag	Cd	In	Sn	Sb	Te	I	Xe
Cs	Ba	*La	Hf	Ta	W	Re	Os	Ir	Pt	Au	Hg	Tl	Pb	Bi	Po	At	Rn
Fr	Ra	†Ac	Rf	Db	Sg	Bh	Hs	Mt	Ds	Rg	Cn	Uut	Uuq	Uup	Uuh	Uus	Uuo

*	Ce	Pr	Nd	Pm	Sm	Eu	Gd	Tb	Dy	Ho	Er	Tm	Yb	Lu
†	Th	Pa	U	Np	Pu	Am	Cm	Bk	Cf	Es	Fm	Md	No	Lr

Ge

The predictive power of Mendeleyev's periodic table was dramatically demonstrated in 1871 when he predicted the existence of a new element whose chemical properties were similar to those of silicon. He called this new element eka-silicon. The German chemist Clemens Winkler discovered this element in 1886 and named it germanium, from the Latin Germania, the name for Germany. Its physical and chemical properties did indeed very closely resemble those of silicon.

Germanium is a dark gray solid with a metallic shine or luster. It is a relatively rare element. It is never found as a pure metal in nature but generally as a mineral combined with oxygen. Its position in the periodic table is halfway between the metals and nonmetals, and it belongs to the class of compounds called metalloids.

Although germanium has metal-like properties, it is a relatively poor conductor of electricity and is therefore called a semiconductor. The addition of small amounts of certain impurities, such as arsenic, gallium, or antimony, to germanium greatly increases its ability to conduct electricity. This process is known as "doping." The doped germanium is used to make transistors that are at the heart of the solid-state electronics industry. With doping, tens of thousands of transistors can now be formed on a small germanium chip, which in effect becomes a small computer. Materials such as germanium have made possible the revolution in electronic miniaturization.

Atomic Number 32

Chemical Symbol Ge

Group IVA—Metalloid

Arsenic

Atomic Number 33

Chemical Symbol As

Group VA—Metalloid

IA																	VIIIA
H	IA											IIIA	IVA	VA	VIA	VIIA	He
Li	Be											B	C	N	O	F	Ne
Na	Mg	IIIB	IVB	VB	VIB	VIIB		VIIIB		IB	IIB	Al	Si	P	S	Cl	Ar
K	Ca	Sc	Ti	V	Cr	Mn	Fe	Co	Ni	Cu	Zn	Ga	Ge	As	Se	Br	Kr
Rb	Sr	Y	Zr	Nb	Mo	Tc	Ru	Rh	Pd	Ag	Cd	In	Sn	Sb	Te	I	Xe
Cs	Ba	*La	Hf	Ta	W	Re	Os	Ir	Pt	Au	Hg	Tl	Pb	Bi	Po	At	Rn
Fr	Ra	†Ac	Rf	Db	Sg	Bh	Hs	Mt	Ds	Rg	Cn	Uut	Uuq	Uup	Uuh	Uus	Uuo

*	Ce	Pr	Nd	Pm	Sm	Eu	Gd	Tb	Dy	Ho	Er	Tm	Yb	Lu
†	Th	Pa	U	Np	Pu	Am	Cm	Bk	Cf	Es	Fm	Md	No	Lr

As

Arsenic is a brittle, crystalline solid at room temperature. Lying between the nonmetals and metals in the periodic table, it is often thought of as a semimetal, or metalloid. For example, it is a poor conductor of electricity, yet it has a steel-gray color. When exposed to air, arsenic quickly tarnishes to a yellow color, eventually turning black.

Although a small amount of arsenic does occur in its free form in nature, most arsenic is found in a number of mineral compounds such as realgar and orpiment. Both are compounds of arsenic and sulfur. The history of arsenic dates back to ancient Greek and Roman times. Its name is probably derived from the Latin name *arsenicum* for orpiment, a common yellow pigment used at the time.

Arsenic is recovered from a mineral such as orpiment, which has a color almost like that of gold, by first roasting it in air to convert the arsenic to an oxide. The oxide is then heated with carbon to remove the oxygen and liberate the arsenic.

In the form of arsenious oxide, a white crystalline powder compound that is also known as white arsenic or arsenic trioxide, arsenic is a well-known poison. It is used as a weed killer and insecticide, being commonly sprayed on fruit to ward off damage by insects. The insecticide called Paris green also contains arsenic and is often sprayed from airplanes over cotton fields to destroy boll-weevil infestations.

Arsenic as a poison has captured the imagination of many a crime writer. The popularity of arsenic as an instrument of murder does have some basis in fact. Before recent advances in autopsy techniques, it was often impossible to detect arsenic in

In Arsenic and Old Lace, *Cary Grant came across two old women who use arsenic to poison their house guests.*

the body of a victim. The victim died with symptoms resembling those of pneumonia. Because arsenic is dangerous, it is one of the elements whose emission into the environment is being monitored and controlled. Although doses as small as 0.1 of a gram of arsenic can be fatal to humans, minute traces can actually stimulate the production of red blood cells.

Although a poison, arsenic compounds have had a history of producing useful medical products. Many skin diseases have been effectively treated with some of these compounds, as has amoebic dysentery. Newer, and potentially less dangerous treatments, have replaced many of these compounds. But perhaps the most celebrated arsenic-containing medicine is compound "606."

Compound 606 was developed by the German scientist Paul Ehrlich in 1910 as a cure for syphilis. Before the days of antibiotics, it was one of the few known treatments for this debilitating disease. Ehrlich named his famous compound 606 because it was the 606th compound he tried during his extensive research in trying to find a cure for syphilis. Its chemical name is arsphenamine.

Arsenic has become a material of great importance in the world of solid-state electronics. Small amounts of arsenic are now added to such semiconductors as germanium and silicon to transform them into transistors. Arsenic also forms a compound with gallium, gallium arsenide (GaAs), that can transform electricity directly into light. This is used to produce light-emitting diodes, or LEDs.

Doses as small as 0.1 of a gram of arsenic can be fatal, and before recent advances in autopsy techniques it was often impossible to detect arsenic in the body of a victim.

Selenium

Atomic Number **34**

Chemical Symbol **Se**

Group **VIA—Metalloid**

IA	IA	IIIB	IVB	VB	VIB	VIIB	VIIIB			IB	IIB	IIIA	IVA	VA	VIA	VIIA	VIIIA
H																	He
Li	Be											B	C	N	O	F	Ne
Na	Mg											Al	Si	P	S	Cl	Ar
K	Ca	Sc	Ti	V	Cr	Mn	Fe	Co	Ni	Cu	Zn	Ga	Ge	As	Se	Br	Kr
Rb	Sr	Y	Zr	Nb	Mo	Tc	Ru	Rh	Pd	Ag	Cd	In	Sn	Sb	Te	I	Xe
Cs	Ba	*La	Hf	Ta	W	Re	Os	Ir	Pt	Au	Hg	Tl	Pb	Bi	Po	At	Rn
Fr	Ra	†Ac	Rf	Db	Sg	Bh	Hs	Mt	Ds	Rg	Cn	Uut	Uuq	Uup	Uuh	Uus	Uuo

*	Ce	Pr	Nd	Pm	Sm	Eu	Gd	Tb	Dy	Ho	Er	Tm	Yb	Lu
†	Th	Pa	U	Np	Pu	Am	Cm	Bk	Cf	Es	Fm	Md	No	Lr

Se

Selenium is a metalloid that can exist in two different forms, red and gray, called allotropes. Red selenium is an amorphous, glasslike solid, whereas gray selenium is a soft, bluish-gray metal. Selenium was discovered by the great Swedish chemist Jöns Jakob Berzelius in 1817. It derives its name from *selene*, the Greek word for the moon.

Selenium-bearing minerals are too scarce to be mined profitably. Because selenium is usually found in the company of copper and sulfur, almost all commercially available selenium is recovered as a by-product of copper refining and the manufacture of sulfuric acid.

Gray selenium is a photoconductor. This means that although it is ordinarily a poor conductor of electricity, it

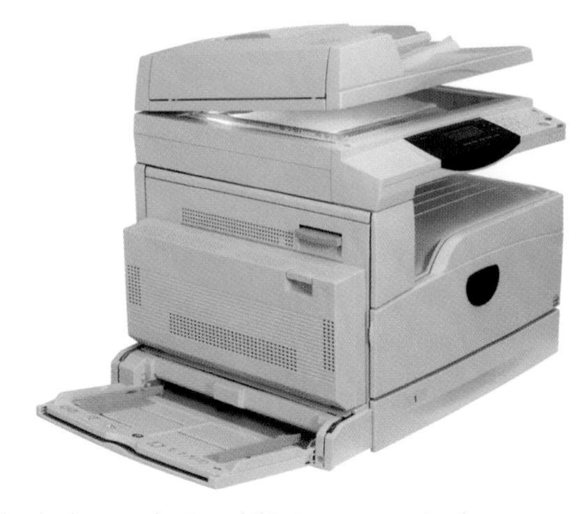

Selenium's photoconducting ability is put to use in photocopy machines.

becomes an excellent conductor in the presence of light. Its ability to conduct electricity can increase by a factor of 1,000 when light falls on it. This makes selenium very valuable as a light sensor. It has found use in such applications as robotics and light-switching devices, as well as light meters.

A major use of the photoconducting ability of selenium is in photocopying. For this purpose a belt or plate is coated with selenium and given an electrostatic charge. It is then exposed to a printed page and light is made to shine through the page. The charge on the lighted parts of the belt is conducted away electrostatically, leaving a charge on the belt only where dark images are present on the original print. A toner, consisting of a finely powdered carbon, is then distributed over the belt and adheres to the electrostatically charged areas. The principle by which the carbon adheres to these areas is similar to the one that makes a rubbed balloon stick to your clothes. Finally, the carbon is fused to a sheet of paper by heat, producing a reproduction of the original document.

Selenium was discovered by Swedish chemist Jöns Jakob Berzelius in 1817. It derives its name from selene, *the Greek word for the moon.*

A great deal of recent interest has also been shown in the effect of selenium on human health. Selenium was recognized as being toxic soon after its discovery. Recent studies have suggested, however, that trace amounts of selenium in the diet can actually protect humans against cancer and heart disease. These results have not been fully confirmed, but it is known that selenium is involved in the activity of vitamin E and certain enzymes.

Compounds containing selenium have also proven useful in controlling dandruff and are often added to shampoos.

Bromine

Atomic Number 35

Chemical Symbol Br

Group VIIA—The Halogens

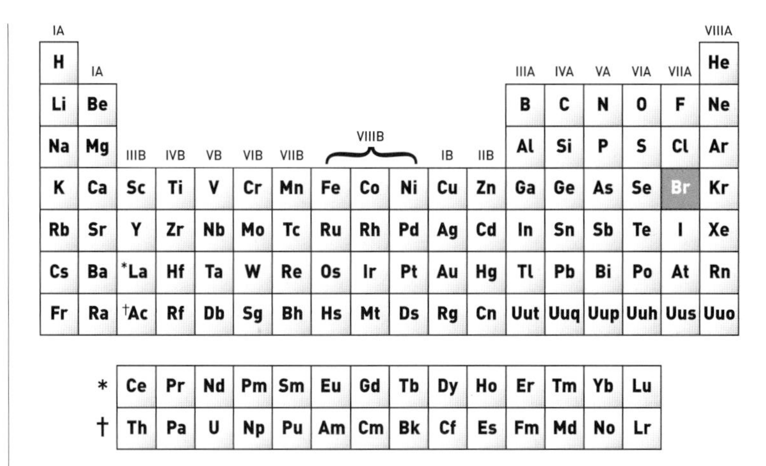

Br

Bromine is a reddish liquid with an acrid smell that produces thick, reddish-brown fumes when exposed to air. The simplest bromine molecule contains two bromine atoms and is called diatomic bromine. Bromine and mercury are the only two elements that are liquids at room temperature. The density of bromine is 3.12 grams/cm^3, making it approximately three times as dense as water. It is, for example, almost impossible to transfer liquid bromine with an ordinary eye dropper because it is too heavy and will leak out. The process of pouring bromine into a beaker is quite striking to watch. Bromine vapor is so much denser than air that it will hug the bottom of the beaker.

Bromine was discovered by a young French student, Antoine-Jérôme Balard, in 1826. He first isolated the element by adding chlorinated water to the brackish water he found in the salt marshes near Montpellier, a city in southern France. The addition of chlorine turned the water brownish. The cause of the color change was found to be a new element that the French Academy named bromine, from the Greek word *bromos*, meaning "stench."

Bromine can be found in seawater, underground salt mines, and deep brine wells. Much of it is taken from the ocean, but the major sources of bromine in the United States are the deep brine wells found in Arkansas, which contain a solution of 0.5 percent bromine. A modification of Balard's method is still used for the commercial production of bromine. Chlorine gas is added to heated brine or seawater. This oxidizes the bromine ions in solution, changing them to elemental bromine, which is then driven out of solution by jets of air or steam.

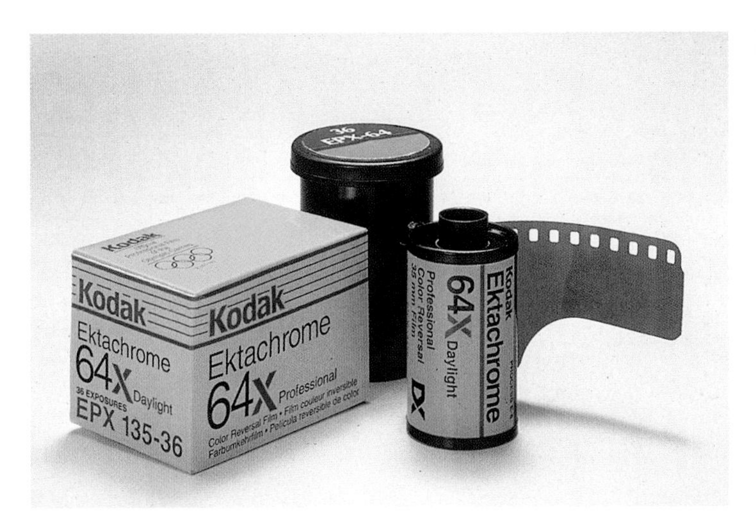

Bromine is used in the manufacture of photographic film. Black-and-white film consists essentially of silver bromide crystals placed in a gelatin to form a light-sensitive emulsion.

A major use of bromine is in producing a gasoline additive called ethylene dibromide. This compound removes lead additives after the combustion of gasoline, preventing the lead in these additives from forming deposits in the engine. Instead, the lead combines with bromine to form lead bromide, a volatile gas, which leaves the engine through the exhaust system. Recent studies have shown that ethylene dibromide is a potent carcinogen.

Methyl bromide is another compound of bromine that has commercial value. It is used as a pesticide and is particularly effective against the parasitic worms known as nematodes. Bromine is also used in the manufacture of silver bromide for photographic film.

Bromine is extremely toxic and can cause severe burns on the skin. Moreover, its vapors are not only irritating and noxious but also can cause severe damage to the tissues of the nose and throat. Sodium bromide, a salt of bromine, is a mild sedative that was widely used in the late 19th and early 20th centuries. People still speak of taking a "bromide" to ease tension.

Bromine is a reddish liquid with an acrid smell. Its name is derived from the Greek word bromos, *meaning "stench."*

Krypton

IA	IA	IIIB	IVB	VB	VIB	VIIB	VIIIB			IB	IIB	IIIA	IVA	VA	VIA	VIIA	VIIIA
H																	He
Li	Be											B	C	N	O	F	Ne
Na	Mg											Al	Si	P	S	Cl	Ar
K	Ca	Sc	Ti	V	Cr	Mn	Fe	Co	Ni	Cu	Zn	Ga	Ge	As	Se	Br	Kr
Rb	Sr	Y	Zr	Nb	Mo	Tc	Ru	Rh	Pd	Ag	Cd	In	Sn	Sb	Te	I	Xe
Cs	Ba	*La	Hf	Ta	W	Re	Os	Ir	Pt	Au	Hg	Tl	Pb	Bi	Po	At	Rn
Fr	Ra	†Ac	Rf	Db	Sg	Bh	Hs	Mt	Ds	Rg	Cn	Uut	Uuq	Uup	Uuh	Uus	Uuo

*	Ce	Pr	Nd	Pm	Sm	Eu	Gd	Tb	Dy	Ho	Er	Tm	Yb	Lu
†	Th	Pa	U	Np	Pu	Am	Cm	Bk	Cf	Es	Fm	Md	No	Lr

Krypton is a noble gas, one of the group of gases in the periodic table that also contains helium and argon; they are called noble gases because of their lack of chemical reactivity. Until recently, no compound of a noble gas was known to exist. The chemical inertness of the noble gases was challenged by the celebrated American chemist Linus Pauling, who in 1933 used quantum mechanics to demonstrate that a molecule containing krypton and fluorine should be stable enough to exist. Finally, in 1966, Neil Bartlett of the University of British Columbia produced the first compound of a noble gas, which, as Pauling had predicted, was a compound of krypton and fluorine. Known as krypton fluoride (KrF_4), it is the only known compound of the element.

Because none of the noble gases had been discovered when Mendeleyev developed his periodic table in 1869, he made no provision for them in the table. It was the discovery of argon in 1894 that demonstrated the need for a new column in the table. The English chemist Sir William Ramsay isolated and identified krypton as an element in 1898. He named it from the Greek word *krytos*, which means "hidden."

In 1933, the American chemist Linus Pauling challenged the idea that the noble gases were chemically inert by predicting the existence of a molecule containing krypton and fluorine. His claim was proved in 1966.

Atomic Number 36

Chemical Symbol Kr

Group VIIIA—The Noble Gases

Krypton exists in trace amounts in the air. Its atmospheric abundance is 0.0001 percent by volume. The commercial preparation of krypton uses essentially the same method used by its discoverer, Ramsay. The gas is separated from liquid air by fractional distillation. It is an odorless, tasteless, colorless, and completely harmless gas. Its chief use is in the "neon" lights that have become an important part of the modern landscape. When sealed in a glass tube and subjected to an electrical discharge, krypton produces a pale violet color and is often used for airport runway and approach lights. It is sometimes mixed with other noble gases to produce subtle variations in their color or brightness. Since it has a fast response time to electric currents, it is often used, mixed with xenon, in high-intensity, short-exposure photographic flash bulbs or strobe lights.

The English chemist Sir William Ramsay isolated and identified krypton as an element in 1898. He named it from the Greek word krytos, *which means "hidden."*

When atoms are excited by the absorption of energy, they often release energy themselves by giving off light. Every element emits light that contains a specific, characteristic series of colors. These colors are called the emission spectrum of the element, and each color has a very precise, measurable wavelength. In 1960, an international committee selected the wavelength of one of the colors of krypton-86 to define the meter. One meter was no longer the distance between two arbitrary scratches on a platinum bar stored in Paris, but was now defined as exactly 1,650,762.73 wavelengths of the red–orange spectral line of krypton-86. The advent of the laser and the ability to measure the speed of light very precisely led, in 1983, to a redefinition of the meter. One meter is now defined so that the speed of light in a vacuum is exactly 299,792,458 meters per second.

Rubidium

IA																	VIIIA
H	IA											IIIA	IVA	VA	VIA	VIIA	He
Li	Be											B	C	N	O	F	Ne
Na	Mg	IIIB	IVB	VB	VIB	VIIB	VIIIB			IB	IIB	Al	Si	P	S	Cl	Ar
K	Ca	Sc	Ti	V	Cr	Mn	Fe	Co	Ni	Cu	Zn	Ga	Ge	As	Se	Br	Kr
Rb	Sr	Y	Zr	Nb	Mo	Tc	Ru	Rh	Pd	Ag	Cd	In	Sn	Sb	Te	I	Xe
Cs	Ba	*La	Hf	Ta	W	Re	Os	Ir	Pt	Au	Hg	Tl	Pb	Bi	Po	At	Rn
Fr	Ra	†Ac	Rf	Db	Sg	Bh	Hs	Mt	Ds	Rg	Cn	Uut	Uuq	Uup	Uuh	Uus	Uuo

*	Ce	Pr	Nd	Pm	Sm	Eu	Gd	Tb	Dy	Ho	Er	Tm	Yb	Lu
†	Th	Pa	U	Np	Pu	Am	Cm	Bk	Cf	Es	Fm	Md	No	Lr

Rb

Rubidium is a very soft, silvery, highly reactive metal that burns spontaneously when exposed to air. It also reacts violently with water, giving off large quantities of hydrogen gas that immediately burst into flames because of the heat generated by the reaction. The pure metal is usually stored in kerosene. Rubidium is much too reactive to exist as a pure metal in nature, and few rubidium-bearing minerals are known.

Rubidium was discovered in 1861 by the German chemists Robert Bunsen and Gustav Kirchhoff. They found it as an impurity among the many alkali metals they were investigating. Bunsen and Kirchhoff identified rubidium by the spectral lines it emitted, one of which has a deep red color. They took its name from the Latin word *rubidus*, meaning "deep red."

Rubidium has little commercial value. It is obtained as a by-product in the refining of such related metals as lithium and cesium, whose ores contain trace amounts of rubidium. The chief use of rubidium is in the manufacture of television and cathode-ray tubes and as a "getter" for vacuum systems. A getter acts as a scavenger and removes unwanted gas that would contaminate the system.

The melting point of rubidium is 39°C. This is only slightly above the human body temperature of 37°C and considerably below the boiling point of water 100°C.

Atomic Number **37**

Chemical Symbol **Rb**

Group **IA—The Alkali Metals**

IA																	VIIIA
H	IA											IIIA	IVA	VA	VIA	VIIA	He
Li	Be											B	C	N	O	F	Ne
Na	Mg	IIIB	IVB	VB	VIB	VIIB	VIIIB			IB	IIB	Al	Si	P	S	Cl	Ar
K	Ca	Sc	Ti	V	Cr	Mn	Fe	Co	Ni	Cu	Zn	Ga	Ge	As	Se	Br	Kr
Rb	Sr	Y	Zr	Nb	Mo	Tc	Ru	Rh	Pd	Ag	Cd	In	Sn	Sb	Te	I	Xe
Cs	Ba	*La	Hf	Ta	W	Re	Os	Ir	Pt	Au	Hg	Tl	Pb	Bi	Po	At	Rn
Fr	Ra	†Ac	Rf	Db	Sg	Bh	Hs	Mt	Ds	Rg	Cn	Uut	Uuq	Uup	Uuh	Uus	Uuo

*	Ce	Pr	Nd	Pm	Sm	Eu	Gd	Tb	Dy	Ho	Er	Tm	Yb	Lu
†	Th	Pa	U	Np	Pu	Am	Cm	Bk	Cf	Es	Fm	Md	No	Lr

Strontium

Atomic Number 38

Chemical Symbol Sr

Group IIA—The Alkaline-Earth Metals

Sr

Strontium is a fairly active, silvery-white metal that quickly tarnishes when exposed to air. It is relatively soft and, when finely divided, will burn spontaneously in air.

The sources of strontium are the mineral ores celestite, which is a form of strontium sulfate, and strontianite, a strontium carbonate. Both strontium and the mineral strontianite were named in 1789 by the English scientist Dr. Adair Crawford, who first identified the element in the Scottish village of Strontian. The metallurgical process for obtaining the pure metal is first to treat its ores with hydrochloric acid, converting their strontium compounds to strontium chloride, and then to use electrolysis to separate the metal from the chloride.

One of the isotopes of strontium, strontium-90, is radioactive, with a rather long half-life of 28.8 years. It is a by-product of nuclear explosions and can often contaminate large areas of the environment through fallout from the atmosphere. Since the chemistry of strontium is similar to that of calcium, strontium-90 can pose a serious health hazard when it is taken up by cows that feed on hay and grass contaminated by fallout because it is incorporated into their milk. In humans who drink the milk, the strontium-90 replaces the calcium in bone tissue, where it can cause radiation damage to the sensitive bone marrow, in which blood cells are formed. Since strontium-90 is produced whenever uranium undergoes fission, the operators of nuclear reactors must be constantly on their guard to prevent its accidental release into the environment.

Strontium metal has little commercial use, and compounds of strontium have found only limited application in industry.

A fireworks display over the Brooklyn Bridge by Fireworks by Grucci. The red color in many fireworks comes from the presence of strontium salts.

Since strontium salts, such as strontium carbonate, give off a characteristic red color when they burn, they are often used in highway warning flares and in fireworks.

The ability of strontium to replace calcium is used in medical research. In one such technique, strontium-87, a radioactive isotope of strontium, is introduced into a patient's body and is taken up by bone tissue. Radiation detectors are then used to locate the radioactive strontium and to assess any abnormalities. The half-life of strontium-97 is only 2.8 hours, so it is rapidly eliminated from the body, keeping exposure to radiation to a minimum.

Yttrium

IA	IA	IIIB	IVB	VB	VIB	VIIB	VIIIB			IB	IIB	IIIA	IVA	VA	VIA	VIIA	VIIIA
H																	He
Li	Be											B	C	N	O	F	Ne
Na	Mg											Al	Si	P	S	Cl	Ar
K	Ca	Sc	Ti	V	Cr	Mn	Fe	Co	Ni	Cu	Zn	Ga	Ge	As	Se	Br	Kr
Rb	Sr	Y	Zr	Nb	Mo	Tc	Ru	Rh	Pd	Ag	Cd	In	Sn	Sb	Te	I	Xe
Cs	Ba	*La	Hf	Ta	W	Re	Os	Ir	Pt	Au	Hg	Tl	Pb	Bi	Po	At	Rn
Fr	Ra	†Ac	Rf	Db	Sg	Bh	Hs	Mt	Ds	Rg	Cn	Uut	Uuq	Uup	Uuh	Uus	Uuo

*	Ce	Pr	Nd	Pm	Sm	Eu	Gd	Tb	Dy	Ho	Er	Tm	Yb	Lu
†	Th	Pa	U	Np	Pu	Am	Cm	Bk	Cf	Es	Fm	Md	No	Lr

Yttrium is found in small quantities in the Earth's crust. Rocks brought back from the moon by the Apollo astronauts, however, had an unexpectedly high yttrium content.

Yttrium owes its strange-sounding name to the Swedish village Ytterby, near Stockholm, where it was first identified in 1789 by Johan Gadolin. It is interesting to note that a quarry located in Ytterby has been the source of several new elements that have been named, in some fashion, for the town. In addition to yttrium, there are ytterbium, terbium, and erbium.

Yttrium is a silvery-gray metal that is active enough to burn spontaneously in air when cut into small pieces. The metal itself has little commercial value, but the oxide of yttrium, a white powder, has found use as the phosphor that produces the red

Scientists were surprised to discover that rocks brought back from the surface of the moon had a very high yttrium content.

Atomic Number **39**

Chemical Symbol **Y**

Group **IIIB—Transition Element**

color for color television tubes. Most yttrium is obtained from monazite sand.

Yttrium has recently also been employed in solid-state lasers. Yttrium aluminum garnet, a compound of yttrium, aluminum, and oxygen (usually referred to as YAG), has the ability to intensify and amplify light energy. As in the case of all substances that exhibit laser activity, the radiation emitted by YAG is concentrated in a very narrow band of wavelengths and is therefore nearly monochromatic (of one color). The YAG laser can operate at high power output and has proven particularly useful in the cutting and drilling of metals.

Metals such as copper have a small but significant resistance to the passage of electricity at room temperatures. This resistance usually results in a waste of energy in the form of heat and in a loss of electrical energy. When their temperature is lowered to only a few degrees above absolute zero, however, almost all metals become superconductors, showing no electrical resistance whatsoever.

Yttrium owes its strange-sounding name to the Swedish village Ytterby, near Stockholm, where it was first identified in 1789 by Johan Gadolin.

Extremely low temperatures are impractical, however, and scientists have tried hard to find materials that would be superconducting at room temperatures. A major breakthrough in this work occurred in 1986 when Karl Alexander Muller and Johannes Georg Bednortz at the IBM laboratories in Zurich, Switzerland, discovered a class of ceramics that showed superconductivity at a temperature of 30°K, or 30° degrees above absolute zero. This still represents a rather cold environment, but it inspired further research to find materials that might become superconducting at still higher temperatures. In 1987, scientists at the University of Alabama and the University of Houston announced the discovery of an yttrium compound, basically an yttrium, copper, and barium oxide, that was superconducting at 93°K. This compound is sometimes called the "1-2-3 compound" because of its 1:2:3 ratio of yttrium to barium to copper. Other mixtures of these elements, called perovskites as a class, are being investigated, and scientists are optimistic that one of them will eventually lead to a practical high-temperature superconductor.

The ability to transmit electrical power without loss of energy would be a great step forward in the development of such devices as trains that ride without friction, levitated above their tracks by magnetic fields. The production of the huge magnetic fields needed for such a task requires enormous electrical currents. The losses caused by electrical resistance in conventional wires at normal temperatures would make the operation of such new technologies too expensive.

IA																	VIIIA
H	IA											IIIA	IVA	VA	VIA	VIIA	He
Li	Be											B	C	N	O	F	Ne
Na	Mg	IIIB	IVB	VB	VIB	VIIB		VIIIB		IB	IIB	Al	Si	P	S	Cl	Ar
K	Ca	Sc	Ti	V	Cr	Mn	Fe	Co	Ni	Cu	Zn	Ga	Ge	As	Se	Br	Kr
Rb	Sr	Y	Zr	Nb	Mo	Tc	Ru	Rh	Pd	Ag	Cd	In	Sn	Sb	Te	I	Xe
Cs	Ba	*La	Hf	Ta	W	Re	Os	Ir	Pt	Au	Hg	Tl	Pb	Bi	Po	At	Rn
Fr	Ra	†Ac	Rf	Db	Sg	Bh	Hs	Mt	Ds	Rg	Cn	Uut	Uuq	Uup	Uuh	Uus	Uuo

*	Ce	Pr	Nd	Pm	Sm	Eu	Gd	Tb	Dy	Ho	Er	Tm	Yb	Lu
†	Th	Pa	U	Np	Pu	Am	Cm	Bk	Cf	Es	Fm	Md	No	Lr

Zirconium

Zr

Zirconium is a strong, durable metal that is resistant to both corrosion and high temperatures. It reacts rapidly with the oxygen in air to form a tough zirconium oxide layer that protects the metal from any further reaction. Zirconium has played a major role in the construction of space-vehicle parts. Its resistance to high temperatures is critical in the design of those parts that are exposed to the extreme heat produced on the re-entry of space vehicles into the Earth's atmosphere.

It is difficult to isolate pure zirconium. Hafnium, the element located immediately below zirconium in the same column of the periodic table, is an inevitable impurity. Since the ions of hafnium and zirconium have the same charge and are approximately the same size, it is not surprising that they have similar properties, which make it difficult to separate the two elements from each other. Zirconium was discovered in 1787 by the German chemist Martin Heinrich Klaproth, whereas hafnium remained undiscovered until 1923, when it was found in zirconium ores.

Probably the best known compound of zirconium is the mineral zircon. This mineral has been known since ancient times and is even referred to in the Bible. Zirconium actually derives its name from *zargun*, which is the Arabic name for zircon and means "gold color."

Zircon is a crystalline compound of zirconium silicate ($ZrSiO_4$). It is found in a wide variety of colors, and when the crystal is cut and polished it is regarded as a semiprecious gem. Zircon has an extremely high index of refraction, which means that it has the ability to bend light through large angles. Because

Atomic Number 40

Chemical Symbol Zr

Group IVB—Transition Element

The space shuttle Atlantis *touches down after a 1992 mission. Zirconium's ability to withstand high temperatures makes it an ideal ingredient for the heat-resistant materials used in spacecraft.*

of this, the colorless crystals of zircon have an unusual brilliance and are sometimes used as a substitute for diamonds.

A major use of zirconium is in nuclear reactors. It is actually Zircaloy, a cheaper alloy of zirconium, that is used. Like zirconium itself, Zircaloy is extremely resistant to corrosion in water. To prevent direct contact of uranium with water, the fuel elements in many reactors usually consist of natural uranium pellets inserted into long Zircaloy tubes, called rods. Hundreds of these fuel rods are then inserted into a coolant, such as water, to absorb and carry away the energy produced by the fission of the uranium.

In certain types of experimental reactors being developed at such research centers as Oak Ridge National Laboratory, in Oak Ridge, Tennessee, the core of the reactor itself is constructed of Zircaloy. It is in the core, a pear-shape, half-inch-thick Zircaloy container, that the fissioning of the uranium occurs.

Although Zircaloy is more expensive than stainless steel, it is used in reactors because it does not readily absorb neutrons. The chain reaction taking place in the reactor depends on neutrons, and any loss of neutrons would seriously interfere with the reactor.

IA	IA	IIIB	IVB	VB	VIB	VIIB	VIIIB			IB	IIB	IIIA	IVA	VA	VIA	VIIA	VIIIA
H																	He
Li	Be											B	C	N	O	F	Ne
Na	Mg											Al	Si	P	S	Cl	Ar
K	Ca	Sc	Ti	V	Cr	Mn	Fe	Co	Ni	Cu	Zn	Ga	Ge	As	Se	Br	Kr
Rb	Sr	Y	Zr	Nb	Mo	Tc	Ru	Rh	Pd	Ag	Cd	In	Sn	Sb	Te	I	Xe
Cs	Ba	*La	Hf	Ta	W	Re	Os	Ir	Pt	Au	Hg	Tl	Pb	Bi	Po	At	Rn
Fr	Ra	†Ac	Rf	Db	Sg	Bh	Hs	Mt	Ds	Rg	Cn	Uut	Uuq	Uup	Uuh	Uus	Uuo

*	Ce	Pr	Nd	Pm	Sm	Eu	Gd	Tb	Dy	Ho	Er	Tm	Yb	Lu
†	Th	Pa	U	Np	Pu	Am	Cm	Bk	Cf	Es	Fm	Md	No	Lr

Niobium

Nb

Niobium is a soft, steel-gray metal that is largely found in the mineral columbite, an ore that also contains iron and manganese. It reacts with the air to form a tough niobium oxide layer that makes the remaining metal resistant to corrosion. The chemical properties of niobium are very similar to those of tantalum, the element that lies directly below it in the periodic table, and the two are always found together. Their chemical similarities make it difficult to isolate pure niobium without contamination by tantalum.

Niobium was discovered by the Englishman Charles Hatchett in 1801 while working with a sample of columbite owned by the British Museum in London. After some confusion caused by its similarity to tantalum, the element was finally named niobium for the mythological goddess Niobe. As a historical note, niobium was originally called columbium, after the mineral in which it was found, and is still occasionally referred to by that name.

Niobium has been important in the history of high-temperature superconductivity. Before the work of Karl Alexander Muller and Johannes Georg Bednortz on high-temperature superconductivity (see yttrium), the scientific world became excited when it was shown that an alloy consisting of a compound of niobium and germanium, called an intermetallic compound, became superconducting at the relatively high temperature of 23.2°K. Equally exciting was that the alloy remained superconducting even with large currents flowing through it. This contrasted with many superconductors, which lose their superconductivity above a certain current range. The ability of

Atomic Number **41**

Chemical Symbol **Nb**

Group **VB—Transition Element**

Because niobium-enhanced steel is capable of withstanding high temperatures for long periods of time, it is used in the construction of nuclear reactors, such as this one in Tennessee.

the alloy to withstand large currents permitted the construction of superconducting magnets for such instruments as the nuclear magnetic resonance scanners used in diagnostic medicine.

Niobium is also added to steel for special purposes. At high temperatures, above about 540°C, the boundaries between the small crystal grains that make up stainless steel undergo a chemical change that weakens them and makes them vulnerable to damage. These weakened boundaries, for example, corrode more easily than the rest of the steel and often have a tendency to crumble under extreme stress. The addition of niobium to the steel prevents this from happening, allowing the steel to withstand much higher temperatures.

Niobium-stabilized stainless steel, as it is called, is often used in the construction of nuclear reactors because of its ability to withstand high temperatures over long periods of time. For the same reason, this special steel is also used in the construction of advanced aircraft, such as those used in the Gemini space program, and in special cutting tools.

IA																	VIIIA
H	IA											IIIA	IVA	VA	VIA	VIIA	He
Li	Be											B	C	N	O	F	Ne
Na	Mg	IIIB	IVB	VB	VIB	VIIB		VIIIB		IB	IIB	Al	Si	P	S	Cl	Ar
K	Ca	Sc	Ti	V	Cr	Mn	Fe	Co	Ni	Cu	Zn	Ga	Ge	As	Se	Br	Kr
Rb	Sr	Y	Zr	Nb	Mo	Tc	Ru	Rh	Pd	Ag	Cd	In	Sn	Sb	Te	I	Xe
Cs	Ba	*La	Hf	Ta	W	Re	Os	Ir	Pt	Au	Hg	Tl	Pb	Bi	Po	At	Rn
Fr	Ra	†Ac	Rf	Db	Sg	Bh	Hs	Mt	Ds	Rg	Cn	Uut	Uuq	Uup	Uuh	Uus	Uuo

*	Ce	Pr	Nd	Pm	Sm	Eu	Gd	Tb	Dy	Ho	Er	Tm	Yb	Lu
†	Th	Pa	U	Np	Pu	Am	Cm	Bk	Cf	Es	Fm	Md	No	Lr

Molybdenum

Mo

Molybdenum is a hard, silvery metal that is mined from an ore called molybdenite (MoS_2). In the United States, fairly large deposits of molybdenite are found in Colorado. In the 18th century, molybdenite was often thought to be an ore of lead. It was the Swedish chemist Carl Wilhelm Scheele who, in 1778, first isolated molybdenum from molybdenite and identified it as a new element. He named it after molybdenite ore.

One of the chief uses of molybdenum is as an additive to steel. In the manufacture of steel from iron, other metals are added to the iron to produce a steel alloy with properties that are particularly well suited for a specific function. Steel containing molybdenum, usually called "moly steel," is well suited for automobile and aircraft-engine parts. It is able to withstand the pressure and temperature changes that are constantly taking place in an engine. For the same reason, it is also used in the manufacture of guns and cannons.

One of the radioactive isotopes of molybdenum, molybdenum-99, is used in hospitals to generate technetium-99, a radioactive isotope that is highly useful for producing pictures of the body's internal organs after being taken internally. The radioactive molybdenum used for this purpose is stored in a fairly small container, usually absorbed into granules of alumina. When the molybdenum-99 isotope decays, it is transformed into technetium-99. When needed, the technetium is taken from the container and administered to the patient.

Atomic Number **42**

Chemical Symbol **Mo**

Group **VIB—Transition Element**

Technetium

IA																	VIIIA
H	IA											IIIA	IVA	VA	VIA	VIIA	He
Li	Be											B	C	N	O	F	Ne
Na	Mg	IIIB	IVB	VB	VIB	VIIB	VIIIB			IB	IIB	Al	Si	P	S	Cl	Ar
K	Ca	Sc	Ti	V	Cr	Mn	Fe	Co	Ni	Cu	Zn	Ga	Ge	As	Se	Br	Kr
Rb	Sr	Y	Zr	Nb	Mo	Tc	Ru	Rh	Pd	Ag	Cd	In	Sn	Sb	Te	I	Xe
Cs	Ba	*La	Hf	Ta	W	Re	Os	Ir	Pt	Au	Hg	Tl	Pb	Bi	Po	At	Rn
Fr	Ra	†Ac	Rf	Db	Sg	Bh	Hs	Mt	Ds	Rg	Cn	Uut	Uuq	Uup	Uuh	Uus	Uuo

*	Ce	Pr	Nd	Pm	Sm	Eu	Gd	Tb	Dy	Ho	Er	Tm	Yb	Lu
†	Th	Pa	U	Np	Pu	Am	Cm	Bk	Cf	Es	Fm	Md	No	Lr

Atomic Number **43**

Chemical Symbol Tc

Group **VIIB—Transition Element**

Tc

Technetium is a most interesting and unusual element. It was the first new element to be produced in the laboratory from another element. Logically, technetium takes its name from the Greek word *teknetos*, which means "artificial." Since every isotope of technetium is radioactive and decays to form an isotope of a new element, it is not surprising that technetium is not found naturally on Earth. However, trace amounts of the element have been found in certain stars.

Technetium was discovered in 1937 by Emilio Segrè and Carlo Perrier. The discovery was actually made in Italy using a sample of material sent to them by the University of California at Berkeley. The sample was made using the cyclotron at Berkeley to bombard the element molybdenum with deuterons. Deuterons are the nuclei of an isotope of hydrogen called deuterium, and they consist of one proton and one neutron. The new element was identified by studying its radioactivity. It is another example of an element whose existence was predicted by the periodic table.

Today technetium is also produced in nuclear reactors. In these reactors, enormous numbers of neutrons are used to bombard molybdenum and produce large quantities of one of the most useful isotopes of technetium, technetium-99m.

When a chemical compound of technetium-99m is injected into the veins of a patient, the isotope will concentrate in certain body organs and its radioactivity will expose a photographic plate, revealing how those organs are functioning. The *m* in technetium-99m means that the isotope is in an unstable form; its nucleus contains more energy than that of the nucleus of "ordinary" technetium-99, which means that it is on the verge of releasing this energy. When it finally does let go and decay, it reverts back to the ordinary form of technetium-99. The energy it releases is in the form of gamma radiation only. A gamma ray is basically a very energetic kind of X ray. Like X rays, gamma rays are extremely penetrating and so can easily be detected outside the body.

In order to assess the condition of certain parts of the body, a doctor usually administers the technetium to the patient along with another chemical that will "bind" it to a certain location in the body. For example, the physician can inject a small amount of tin along with the technetium-99m. In the presence of tin, the technetium binds very strongly to red blood cells. On the other hand, in the presence of a phosphorus compound, called a pyrophosphate, the technetium will bind very strongly to the heart muscle. In each case, however, the gamma radiation given off by the technetium-99m can be detected by special equipment and provides an image of the heart so that the doctor can assess the damage done during a heart attack.

Technetium was discovered in Italy in 1937 by Emilio Segrè and Carlo Perrier using a sample of material sent to them by scientists at the University of California at Berkeley.

The half-life of technetium-99m is only 6.02 hours, which makes it an ideal radioisotope for this procedure. It allows measurements to be taken, yet within 24 hours will be almost completely gone from the body. Ordinary technetium-99, the material to which technetium-99m decays, has a long half-life of 212,000 years, so that in the small doses given a patient, its activity does not present a problem. It is quickly eliminated from the body by biological means.

There are actually 19 known isotopes of technetium, and many of them have extremely long half-lives. Technetium-97 and technetium-98, for example, have half-lives of several million years. A long half-life means that an isotope is only weakly radioactive and that very few of its atoms disintegrate within a fixed period of time.

Although difficult to fabricate and rather expensive, some gram-size samples of metallic technetium have been made. It has a silver color when freshly prepared but slowly tarnishes when exposed to air. Trace quantities of the longer-lived isotopes are often added to steel to inhibit corrosion, but these special steels are used sparingly because of their radioactivity.

Ruthenium

IA	IA	IIIB	IVB	VB	VIB	VIIB	VIIIB			IB	IIB	IIIA	IVA	VA	VIA	VIIA	VIIIA
H																	He
Li	Be											B	C	N	O	F	Ne
Na	Mg											Al	Si	P	S	Cl	Ar
K	Ca	Sc	Ti	V	Cr	Mn	Fe	Co	Ni	Cu	Zn	Ga	Ge	As	Se	Br	Kr
Rb	Sr	Y	Zr	Nb	Mo	Tc	Ru	Rh	Pd	Ag	Cd	In	Sn	Sb	Te	I	Xe
Cs	Ba	*La	Hf	Ta	W	Re	Os	Ir	Pt	Au	Hg	Tl	Pb	Bi	Po	At	Rn
Fr	Ra	†Ac	Rf	Db	Sg	Bh	Hs	Mt	Ds	Rg	Cn	Uut	Uuq	Uup	Uuh	Uus	Uuo

*	Ce	Pr	Nd	Pm	Sm	Eu	Gd	Tb	Dy	Ho	Er	Tm	Yb	Lu
†	Th	Pa	U	Np	Pu	Am	Cm	Bk	Cf	Es	Fm	Md	No	Lr

Ru

Ruthenium is a rare element that is usually recovered as a by-product of the refining of platinum ores. It is a hard, brittle, silvery metal with very little commercial value. Its discovery is usually attributed to the Estonian chemist Karl Karlovitch Klaus, who named it from the Latin word *Ruthenia*, meaning "Russia."

Ruthenium is mainly used as a catalyst for many types of industrial processes. A catalyst hastens certain chemical reactions without being changed itself. A recent example of the use of ruthenium as a catalyst is in obtaining hydrogen gas by directly splitting water molecules rather than by the usual process of electrolysis, which uses electric current. The catalytic method depends on energetically exciting ruthenium by exposing it to solar radiation. In this excited state it is capable of separating the hydrogen from the water molecule to form molecular hydrogen.

Ruthenium is also used in the jewelry industry as a hardening additive for platinum and is often added to titanium to improve its resistance to corrosion. Other alloys of ruthenium are used for fountain pen points and special electrical contacts.

Atomic Number **44**

Chemical Symbol **Ru**

Group **VIIIB—Transition Element**

IA																	VIIIA
H	IA											IIIA	IVA	VA	VIA	VIIA	He
Li	Be											B	C	N	O	F	Ne
Na	Mg	IIIB	IVB	VB	VIB	VIIB	VIIIB			IB	IIB	Al	Si	P	S	Cl	Ar
K	Ca	Sc	Ti	V	Cr	Mn	Fe	Co	Ni	Cu	Zn	Ga	Ge	As	Se	Br	Kr
Rb	Sr	Y	Zr	Nb	Mo	Tc	Ru	Rh	Pd	Ag	Cd	In	Sn	Sb	Te	I	Xe
Cs	Ba	*La	Hf	Ta	W	Re	Os	Ir	Pt	Au	Hg	Tl	Pb	Bi	Po	At	Rn
Fr	Ra	†Ac	Rf	Db	Sg	Bh	Hs	Mt	Ds	Rg	Cn	Uut	Uuq	Uup	Uuh	Uus	Uuo

*	Ce	Pr	Nd	Pm	Sm	Eu	Gd	Tb	Dy	Ho	Er	Tm	Yb	Lu
†	Th	Pa	U	Np	Pu	Am	Cm	Bk	Cf	Es	Fm	Md	No	Lr

Rhodium

Rh

Rhodium is an extremely hard, corrosion-resistant, silvery-gray metal. It is quite rare and is usually found associated with the ores of platinum. It was discovered by the Englishman William Hyde Wollaston in 1803. He named it after the Greek word *rhodon* for "rose" because many of the salts of rhodium have a striking rose color.

Rhodium has a number of commercial applications, such as an additive for hardening platinum, a catalyst for some industrial processes, an alloying agent, and a component in aircraft turbine engines. The automotive industry, however, is the major consumer of rhodium. There it is used in the construction of catalytic converters. The exhaust gases produced by the internal-combustion engines of automobiles contain nitrogen dioxide, carbon monoxide, and various unburned hydrocarbons, all of which are major sources of air pollution. The catalytic converter of an automobile is filled with small catalytic beads containing platinum, palladium, and rhodium, which convert the hot exhaust gases that pass through them into harmless products. The carbon monoxide and unburned hydrocarbons are converted to carbon dioxide and water, whereas the nitrogen dioxide is converted to nitrogen and oxygen. Another use for rhodium in automobiles is in the construction of headlight reflectors.

Although a nontarnishing precious metal, rhodium has limited use in the making of jewelry. Its high melting point makes it difficult to cast, and its high cost, often greater than gold or platinum, restrict its use to plating such artifacts as white gold pendants and wedding rings to improve their "shine."

Atomic Number **45**

Chemical Symbol **Rh**

Group **VIIIB—Transition Element**

Palladium

Atomic Number **46**

Chemical Symbol **Pd**

Group **VIIIB—Transition Element**

IA	IA	IIIB	IVB	VB	VIB	VIIB	VIIIB			IB	IIB	IIIA	IVA	VA	VIA	VIIA	VIIIA
H																	He
Li	Be											B	C	N	O	F	Ne
Na	Mg											Al	Si	P	S	Cl	Ar
K	Ca	Sc	Ti	V	Cr	Mn	Fe	Co	Ni	Cu	Zn	Ga	Ge	As	Se	Br	Kr
Rb	Sr	Y	Zr	Nb	Mo	Tc	Ru	Rh	Pd	Ag	Cd	In	Sn	Sb	Te	I	Xe
Cs	Ba	*La	Hf	Ta	W	Re	Os	Ir	Pt	Au	Hg	Tl	Pb	Bi	Po	At	Rn
Fr	Ra	†Ac	Rf	Db	Sg	Bh	Hs	Mt	Ds	Rg	Cn	Uut	Uuq	Uup	Uuh	Uus	Uuo

*	Ce	Pr	Nd	Pm	Sm	Eu	Gd	Tb	Dy	Ho	Er	Tm	Yb	Lu
†	Th	Pa	U	Np	Pu	Am	Cm	Bk	Cf	Es	Fm	Md	No	Lr

Palladium is a quite soft, silvery-white metal that resembles platinum. It is extremely malleable and ductile, meaning it can be hammered into thin sheets and drawn into fine wires. It is relatively rare and, like rhodium and ruthenium, is often found associated with platinum. Like rhodium, it was discovered in 1803 by William Hyde Wollaston, who named it after the asteroid Pallas.

Palladium is an excellent catalyst that was found to greatly facilitate carbon–carbon bonding in organic molecules. This has made possible the making of a host of new compounds in medicine, agriculture, and electronics. Three chemists, Richard Heck, Ei-ichi Negishi, and Akira Suzuki, shared the Nobel Prize in Chemistry in 2010 for the discovery.

Metallic hydrides are formed when palladium is treated with hydrogen gas. Under these conditions, the hydrogen molecules break apart at the surface of the metal and migrate into the metal to occupy the holes in its crystal structure. Palladium can absorb more than 900 times its volume of hydrogen gas. This phenomenon is often used to purify hydrogen. The gas is placed in a container with a very thin palladium wall. The hydrogen gas diffuses into and then through the wall, which traps impurities.

Certain palladium compounds have been found to be effective agents in the treatment of cancerous tumors. These compounds inhibit cell division, and many are free of troublesome side effects. A radioactive isotope, palladium-103, is also useful in the treatment of prostate cancer. Introduced in the body in the form of a small pellet, it can destroy the cancer, and with a half-life of only 17 days it can quickly disappear from the body.

Palladium is extremely resistant to corrosion and is often used in dentistry and the making of jewelry.

IA	IA	IIIB	IVB	VB	VIB	VIIB	VIIIB			IB	IIB	IIIA	IVA	VA	VIA	VIIA	VIIIA
H																	He
Li	Be											B	C	N	O	F	Ne
Na	Mg											Al	Si	P	S	Cl	Ar
K	Ca	Sc	Ti	V	Cr	Mn	Fe	Co	Ni	Cu	Zn	Ga	Ge	As	Se	Br	Kr
Rb	Sr	Y	Zr	Nb	Mo	Tc	Ru	Rh	Pd	Ag	Cd	In	Sn	Sb	Te	I	Xe
Cs	Ba	*La	Hf	Ta	W	Re	Os	Ir	Pt	Au	Hg	Tl	Pb	Bi	Po	At	Rn
Fr	Ra	†Ac	Rf	Db	Sg	Bh	Hs	Mt	Ds	Rg	Cn	Uut	Uuq	Uup	Uuh	Uus	Uuo

*	Ce	Pr	Nd	Pm	Sm	Eu	Gd	Tb	Dy	Ho	Er	Tm	Yb	Lu
†	Th	Pa	U	Np	Pu	Am	Cm	Bk	Cf	Es	Fm	Md	No	Lr

Silver

Atomic Number **47**

Chemical Symbol **Ag**

Group **IB—Transition Element (Coinage Metal)**

Ag

Silver is one of the few metals found in its free state in nature. It is also found in fairly high concentrations in such ores as argentite, a silver sulfide, and as an impurity in copper–nickel ores. Silver was known in ancient times and derives its chemical symbol, Ag, from the Latin word *argentum*, which means silver. The alchemists had a special symbol for it, consisting of a sign in the shape of a new moon. Silver has been used for money since biblical times and so is often referred to as a coinage metal. Its "silver" color and luster are well known and are used to describe the appearance of other metals. Of all the metals, silver is the best conductor of heat and electricity. It is not usually used in home wiring because of its expense, but it is extensively used in the manufacture of high-quality electronic equipment.

Silver has been known and used since ancient times. An 18th-century French engraving shows silversmiths at work making jewelry and flatware.

The sterling silver used to create these eating utensils is a mix of 7 percent copper and 93 percent silver.

Silver is a soft metal that is extremely ductile and malleable. This means that it can be drawn into fine wires and hammered into extremely thin sheets for a variety of uses. In making jewelry, copper is usually alloyed with silver to harden the latter and give it the strength needed. Sterling silver, for example, is 7 percent copper, whereas some so-called silver rings and bracelets can contain up to 20 percent copper. The amount of silver contained in jewelry is often expressed in terms of "fineness." This is simply 10 times the percentage of silver contained in a piece of jewelry. Sterling silver, which is 93 percent silver, thus has a fineness of 930. Much of European silverware has a fineness of 800. When polished, silver can attain an exceptionally high luster, which makes it valuable as a reflective coating on mirrors.

Silver is often used to plate jewelry and eating utensils. In the plating process, an electrical current is used to deposit a thin, protective coating of silver onto less expensive metals, which greatly improves their appearance and durability. Silver electroplating is usually accomplished by connecting the electrically positive terminal, or anode, of a power source to a bar of silver, connecting the negative terminal of the source to the object to be plated, and immersing both in a bath of silver cyanide. As electrical current flows through this miniature battery cell, a bright, shiny film of silver forms on the object.

Metallic silver does not react with the oxygen in air, but it does slowly tarnish over time. The black film that forms on silver objects is silver sulfide. The decay of vegetable matter always results in the presence of small quantities of hydrogen sulfide in the air, and it is this pollutant that reacts with the silver to produce a tarnish. The tarnishing process can be very rapid if a silver utensil is left in contact with a food that contains sulfur, such as eggs or mustard.

Silver is used very extensively in a variety of commercial and scientific products. Industries in the United States use more than 175 million ounces of silver every year, with about 24 percent of this quantity used by the electronics industry. Much of the silver consumed in the United States is also used to make photographic film and paper.

One of the more common uses of silver is in the tooth fillings used in dentistry. A typical filling is a mixture of silver powder and mercury. The silver, usually added in excess amounts, dissolves in the mercury to form an amalgam. The amalgam is actually a solution of silver in mercury, which has the effect of cementing the silver particles together.

Silver compounds are very sensitive to light. Many silver salts, such as silver chloride and silver iodide, darken on exposure to light. The darkening process, called photochemical decomposition, is caused by light energy decomposing the silver salt to form metallic silver. In photography, the emulsion used for photographic film and paper usually consists of a gelatin mixed with silver bromide. When you take a snapshot, light falling on the film energizes the silver ions in the molecules of the silver bromide. These "excited" silver ions then form silver atoms on the surface of the emulsion, producing a latent image of whatever was being photographed. A developing material is then added to "fix" this image.

Photochromic sunglasses, which darken in sunlight, also depend upon the decomposition of silver salts by light. In these glasses, small grains of silver chloride are mixed with the glass used for the lenses. In the presence of light, a thin layer of opaque metallic silver forms on the surface of the glass. The photochemical decomposition that causes this to happen is made reversible by the addition of copper ions to the glass. In the absence of light, the copper ions change the silver atoms back to their original form once again, yielding silver chloride.

Much of the silver consumed in the United States is used for industrial purposes.

Crystals of silver iodide, with their yellowish color, have found an interesting application outside of photography. They are used to "seed" clouds and produce rain. When dropped from an airplane, these tiny crystals cause water drops to form around them, which then enlarge until the weight of each drop is sufficient to make it fall to Earth as rain. The silver iodide in the rain is relatively harmless to plants and animals and is dispersed by the rain itself.

Like gold, silver has become an important element in the field of nanotechnology. Silver nanoparticles are typically between 1 and 100 nanometers (billionths of a meter) in size. Only slightly larger than atoms and molecules, their chemical behavior and their interaction with light often differ from silver in bulk. The color of these tiny silver particles, for example, is yellow, a fact already known hundreds of years ago to artisans who made yellow stained glass in medieval cathedrals. Among their many industrial applications are inks for printed electronic circuits, conducting sheets, and coatings. Their strong antibacterial properties have led to many applications: bandages can be treated with these nanoparticles to prevent infection, the packaging of foods coated to extend shelf life, and socks treated to inhibit the growth of odor-producing bacteria.

With the rapid growth of applications for silver nanoparticles (estimated to be used in more than 150 consumer products), many scientists have become concerned that the toxic nature of these particles might damage the environment by destroying the activity of useful, benign bacteria in waste and sewage systems.

Cadmium

IA																	VIIIA
H	IA											IIIA	IVA	VA	VIA	VIIA	He
Li	Be											B	C	N	O	F	Ne
Na	Mg	IIIB	IVB	VB	VIB	VIIB		VIIIB		IB	IIB	Al	Si	P	S	Cl	Ar
K	Ca	Sc	Ti	V	Cr	Mn	Fe	Co	Ni	Cu	Zn	Ga	Ge	As	Se	Br	Kr
Rb	Sr	Y	Zr	Nb	Mo	Tc	Ru	Rh	Pd	Ag	Cd	In	Sn	Sb	Te	I	Xe
Cs	Ba	*La	Hf	Ta	W	Re	Os	Ir	Pt	Au	Hg	Tl	Pb	Bi	Po	At	Rn
Fr	Ra	†Ac	Rf	Db	Sg	Bh	Hs	Mt	Ds	Rg	Cn	Uut	Uuq	Uup	Uuh	Uus	Uuo

*	Ce	Pr	Nd	Pm	Sm	Eu	Gd	Tb	Dy	Ho	Er	Tm	Yb	Lu
†	Th	Pa	U	Np	Pu	Am	Cm	Bk	Cf	Es	Fm	Md	No	Lr

Atomic Number 48

Chemical Symbol Cd

Group IIB—Transition Element

Cd

Cadmium is a silvery metal that is soft enough to cut with a knife. Its chemical properties greatly resemble those of zinc. It is a relatively rare element, with an abundance of only about 0.001 that of zinc in the Earth's crust. Its principal ore is cadmium sulfide, but cadmium is present in such great quantities in zinc ores that it is generally collected as a by-product of zinc refining. Even its history is intertwined with that of zinc. It was discovered in 1817 by Friedrich Strohmeyer, a pharmaceutical chemist who was investigating the compound zinc oxide, known as calamine.

The major use of cadmium is in the electroplating of steel to protect it from corrosion. It is used less often than zinc for this purpose because it is less abundant than zinc and therefore more expensive and can cause a variety of health problems, including kidney failure and high blood pressure. However, it is much more durable than zinc in an alkaline environment. Only a very thin film of cadmium, about 0.05 millimeters, is required for the electroplating of steel.

Until quite recently the pollution of lakes and rivers by cadmium waste discharges from electroplating industries was a major problem. Tobacco leaves also contain some cadmium, so both smokers and the smoke they exhale are likely to contain trace quantities of the element.

In addition to its use in electroplating, cadmium is used to make nickel–cadmium, or nicad, batteries. The nicad battery can be recharged an indefinite number of times and is therefore much more convenient to use than the rechargeable lead storage battery. It is also much more portable because it can be sealed

The discharge of cadmium waste into lakes and rivers by electroplating companies was, until recently, a major source of pollution.

Tobacco leaves also contain some cadmium, so both smokers and the smoke they exhale are likely to contain trace quantities of the element.

and manufactured in a form very similar to the common dry cell. It produces an output of 1.4 volts, which is slightly less than the 1.5 volts of a common dry cell.

The ability of cadmium to absorb neutrons has made it of great importance in the design of nuclear reactors. Many reactors use cadmium in their control rods. During normal operation of the reactor, these rods are withdrawn from the containment vessel. To stop the reaction, the rods are lowered into the vessel, where they absorb enough neutrons to prevent the self-sustaining chain reaction from continuing.

An alloy called Wood's metal consists chiefly of bismuth but contains about 12.5 percent cadmium. It has an extremely low melting point of 70°C and is used to seal the overhead sprinkler systems installed in many factories and homes. A fire will melt the seal before it gets too hot, releasing water to quench the fire. Cadmium is also used as a red or yellow pigment in making paint.

Indium

IA	IA	IIIB	IVB	VB	VIB	VIIB	VIIIB			IB	IIB	IIIA	IVA	VA	VIA	VIIA	VIIIA
H																	He
Li	Be											B	C	N	O	F	Ne
Na	Mg											Al	Si	P	S	Cl	Ar
K	Ca	Sc	Ti	V	Cr	Mn	Fe	Co	Ni	Cu	Zn	Ga	Ge	As	Se	Br	Kr
Rb	Sr	Y	Zr	Nb	Mo	Tc	Ru	Rh	Pd	Ag	Cd	In	Sn	Sb	Te	I	Xe
Cs	Ba	*La	Hf	Ta	W	Re	Os	Ir	Pt	Au	Hg	Tl	Pb	Bi	Po	At	Rn
Fr	Ra	†Ac	Rf	Db	Sg	Bh	Hs	Mt	Ds	Rg	Cn	Uut	Uuq	Uup	Uuh	Uus	Uuo

*	Ce	Pr	Nd	Pm	Sm	Eu	Gd	Tb	Dy	Ho	Er	Tm	Yb	Lu
†	Th	Pa	U	Np	Pu	Am	Cm	Bk	Cf	Es	Fm	Md	No	Lr

In

Indium is a rare, soft, bluish-white metal that is extremely corrosion-resistant. It is soft enough to leave traces of itself when it is vigorously rubbed against other metals. When bent rapidly, indium gives off an unusual "shrieking" sound. Like several of the elements that precede it in the periodic table, the presence of indium is linked to that of zinc, and most of the world's supply of this element comes as a by-product of the refining of zinc.

Indium was discovered in 1863 by the German chemist Ferdinand Reich, who was investigating elements associated with zinc ore. It was identified by the bright violet light it emitted during spectroscopic analysis. Its name is, in fact, taken from the Latin word *indicum*, which means "violet" or "indigo."

Pure indium metal has few commercial applications and is chiefly used in alloys with other metals. Indium is often used to lower the melting point of the metals with which it is alloyed. Alloys of indium and silver or indium and lead are better conductors of heat than silver or lead alone. They have also found use in the manufacture of transistors and photo cells.

Indium foils are often inserted into nuclear reactors to control the nuclear reaction. The rate at which these foils become radioactive when interacting with the neutrons in the reactor can serve as a valuable measurement of the reactions taking place in the reactor.

Atomic Number 49

Chemical Symbol In

Group IIIA—Post-Transition Metal

Tin

IA	IA	IIIB	IVB	VB	VIB	VIIB	VIIIB	VIIIB	VIIIB	IB	IIB	IIIA	IVA	VA	VIA	VIIA	VIIIA
H																	He
Li	Be											B	C	N	O	F	Ne
Na	Mg											Al	Si	P	S	Cl	Ar
K	Ca	Sc	Ti	V	Cr	Mn	Fe	Co	Ni	Cu	Zn	Ga	Ge	As	Se	Br	Kr
Rb	Sr	Y	Zr	Nb	Mo	Tc	Ru	Rh	Pd	Ag	Cd	In	Sn	Sb	Te	I	Xe
Cs	Ba	*La	Hf	Ta	W	Re	Os	Ir	Pt	Au	Hg	Tl	Pb	Bi	Po	At	Rn
Fr	Ra	†Ac	Rf	Db	Sg	Bh	Hs	Mt	Ds	Rg	Cn	Uut	Uuq	Uup	Uuh	Uus	Uuo

*	Ce	Pr	Nd	Pm	Sm	Eu	Gd	Tb	Dy	Ho	Er	Tm	Yb	Lu
†	Th	Pa	U	Np	Pu	Am	Cm	Bk	Cf	Es	Fm	Md	No	Lr

Sn

Tin is a relatively rare element, ranking 50th or so in the abundance of elements in the Earth's crust. Much of it is found concentrated in deposits of the tin ore cassiterite (tin oxide), so that fairly large quantities of tin are available. The metal can be easily recovered from the ore by heating the latter in the presence of carbon. Tin was among the first metals to be used by human beings. Bronze, an alloy of 80 percent copper and 20 percent tin, was made in Egypt as long ago as 3500 B.C.

Tin is thought to take its name from the Etruscan god Tinia and its chemical symbol from *stannum*, the Latin word for the element. The alchemists worked with tin and created a special symbol for it that looks like the written number 4.

Tin is interesting because it has three different natural forms, called allotropes by chemists. The first is a typically metallic, malleable form called white tin, and the second is a brittle, powdery allotrope called gray tin. White tin is stable above 13.2°C, but at temperatures below this it changes

A pewter chalice from the 18th century. Alloys of tin such as pewter and bronze were once widely used to make household utensils.

Atomic Number **50**

Chemical Symbol **Sn**

Group **IVA—Post-Transition Metal**

Tin is thought to take its name from the Etruscan god Tinia and its chemical symbol from stannum, *the Latin word for the element.*

slowly to gray tin. The lower the temperature the more rapid the change. The transition from white to gray tin was very dramatically illustrated in 1850. During a particularly cold winter in Russia that year, record low temperatures persisted for a long time. The extreme cold not only caused the tin buttons worn by the Russian soldiers to crumble, but also many of the tin pipes used for church organs crumbled. The problem came to be called the "tin disease." The third allotrope of tin, called brittle tin, exists above 161°C. Its properties are obvious from its name.

Tin is too soft a metal to be of much value as a building material or a component of tools. It is very malleable and can be rolled into thin sheets of the tin foil that has been used for hundreds of years. At one time it was quite popular as a material for household utensils such as tin plates and cups. Today it is chiefly used as an alloying agent and to make tin plate, which is steel sheeting covered with a thin coating of tin that is applied electrolytically. Because the tin protects the steel from attack by food acids, tin plate was once used to make tin cans for food but has since been largely replaced by plastic and aluminum. As an alloying metal, tin is used with copper to make bronze and is used to make solder (33 percent tin and 67 percent lead), an easily meltable metal used with a heating iron to make secure electrical connections. Pewter (85 percent tin, 7 percent copper, 6 percent bismuth, and 2 percent antimony), another alloy of tin, has long been used for making cutlery, bowls, and plates. The alloy known as Wood's metal contains about 12.5 percent tin and is used in fire sprinklers, described in the section on cadmium.

Another alloy of tin with lead and antimony is the metal used for casting printing type. Both the tin salts stannic chloride ($SnCl_4$) and stannic oxide (SnO_2) are used to produce glazes for ceramics and special coatings for fabrics.

IA																	VIIIA
H	IA											IIIA	IVA	VA	VIA	VIIA	He
Li	Be											B	C	N	O	F	Ne
Na	Mg	IIIB	IVB	VB	VIB	VIIB	VIIIB			IB	IIB	Al	Si	P	S	Cl	Ar
K	Ca	Sc	Ti	V	Cr	Mn	Fe	Co	Ni	Cu	Zn	Ga	Ge	As	Se	Br	Kr
Rb	Sr	Y	Zr	Nb	Mo	Tc	Ru	Rh	Pd	Ag	Cd	In	Sn	Sb	Te	I	Xe
Cs	Ba	*La	Hf	Ta	W	Re	Os	Ir	Pt	Au	Hg	Tl	Pb	Bi	Po	At	Rn
Fr	Ra	†Ac	Rf	Db	Sg	Bh	Hs	Mt	Ds	Rg	Cn	Uut	Uuq	Uup	Uuh	Uus	Uuo

*	Ce	Pr	Nd	Pm	Sm	Eu	Gd	Tb	Dy	Ho	Er	Tm	Yb	Lu
†	Th	Pa	U	Np	Pu	Am	Cm	Bk	Cf	Es	Fm	Md	No	Lr

Antimony

Atomic Number **51**

Chemical Symbol **Sb**

Group VA—Metalloid

Sb

In its pure form, the metal antimony is a hard, brittle, grayish crystalline solid at room temperature. Although known as a metal, it is a poor conductor of electricity. The ore that serves as the primary source of antimony is the mineral stibnite (antimony sulfide), which has been known for thousands of years. A black compound, stibnite was used in ancient times to darken women's eyebrows. The alchemists often experimented with stibnite, mistaking for lead the metal that it produced upon heating.

The chemical symbol for antimony, Sb, was taken from its original name, stibium, apparently named after the mineral stibnite. The word *antimony* is thought to be of Greek origin.

A major use of antimony is for the common safety match. An item that is now taken for granted, the safety match was invented by the Swedish chemist Jerry Eugene Lundstrom in 1855. The head of this match consists of a mixture of antimony trisulfide and an oxidizing agent such as potassium chlorate. The tip of the match, above the head, contains red phosphorus that, when struck against a rough surface, ignites and generates enough heat to set fire to the head of the match, which bursts into flame.

Antimony oxide, a flame retardant, is often added to the plastic used to make credit cards.

Antimony has few other commercial uses. As an alloy, it can increase the hardness of many metals. Antimony oxide, a white salt, is often added to polyvinyl chloride, or PVC, in which it acts as a flame retardant. PVC is a plastic polymer that is used to make wastewater pipes, credit cards, and electrical insulation.

Tellurium

IA	IA	IIIB	IVB	VB	VIB	VIIB	VIIIB			IB	IIB	IIIA	IVA	VA	VIA	VIIA	VIIIA
H																	He
Li	Be											B	C	N	O	F	Ne
Na	Mg											Al	Si	P	S	Cl	Ar
K	Ca	Sc	Ti	V	Cr	Mn	Fe	Co	Ni	Cu	Zn	Ga	Ge	As	Se	Br	Kr
Rb	Sr	Y	Zr	Nb	Mo	Tc	Ru	Rh	Pd	Ag	Cd	In	Sn	Sb	**Te**	I	Xe
Cs	Ba	*La	Hf	Ta	W	Re	Os	Ir	Pt	Au	Hg	Tl	Pb	Bi	Po	At	Rn
Fr	Ra	†Ac	Rf	Db	Sg	Bh	Hs	Mt	Ds	Rg	Cn	Uut	Uuq	Uup	Uuh	Uus	Uuo

*	Ce	Pr	Nd	Pm	Sm	Eu	Gd	Tb	Dy	Ho	Er	Tm	Yb	Lu
†	Th	Pa	U	Np	Pu	Am	Cm	Bk	Cf	Es	Fm	Md	No	Lr

Te

Tellurium is a rare, silvery-white metalloid belonging to the family of elements that includes such nonmetallic elements as oxygen and sulfur. It is not surprising then that tellurium does not behave like a typical metal. It is very brittle, for example, and does not conduct electricity very well. And just as sulfur and oxygen can combine with other metals to form sulfides, tellurium combines with gold to form a telluride.

Tellurium is one of the few elements that combines with gold. The compounds it forms are called gold tellurides, and they make up a very important component of gold-bearing ores. As such, tellurium is often recovered as a by-product in the refining of gold. It is also found associated with copper-bearing ores and is therefore recovered as a by-product of the refining of copper as well.

The element was discovered in Germany in 1782 by an inspector of mines, Franz Joseph von Reichenstein, an amateur scientist, who was experimenting with ore he found in a gold mine in Transylvania. It was later named in Berlin from the Latin word *tellus*, meaning earth.

The chief use of tellurium is as an additive to such metals as copper and stainless steel to create an alloy that is easier to machine than the original metal.

Atomic Number **52**

Chemical Symbol **Te**

Group **VIA—Metalloid**

IA	IA																VIIIA
H												IIIA	IVA	VA	VIA	VIIA	He
Li	Be											B	C	N	O	F	Ne
Na	Mg	IIIB	IVB	VB	VIB	VIIB	VIIIB			IB	IIB	Al	Si	P	S	Cl	Ar
K	Ca	Sc	Ti	V	Cr	Mn	Fe	Co	Ni	Cu	Zn	Ga	Ge	As	Se	Br	Kr
Rb	Sr	Y	Zr	Nb	Mo	Tc	Ru	Rh	Pd	Ag	Cd	In	Sn	Sb	Te	I	Xe
Cs	Ba	*La	Hf	Ta	W	Re	Os	Ir	Pt	Au	Hg	Tl	Pb	Bi	Po	At	Rn
Fr	Ra	†Ac	Rf	Db	Sg	Bh	Hs	Mt	Ds	Rg	Cn	Uut	Uuq	Uup	Uuh	Uus	Uuo

*	Ce	Pr	Nd	Pm	Sm	Eu	Gd	Tb	Dy	Ho	Er	Tm	Yb	Lu
†	Th	Pa	U	Np	Pu	Am	Cm	Bk	Cf	Es	Fm	Md	No	Lr

Iodine

Atomic Number 53

Chemical Symbol I

Group VIIA—The Halogens

I

Iodine is a violet–black solid that vaporizes to give a violet gas. It is found in seaweed and brine wells, as well as in the sea, in the form of inorganic salts and organic iodides. It is the heaviest element in the halogen family and is highly reactive. Iodine was discovered in 1811 by Bernard Courtois, who noticed a purple vapor emanating from a sample of kelp to which he had added sulfuric acid. The color of the gas has determined its name, which is taken from the Greek word *iodes*, meaning "violet-colored."

Commercial quantities of iodine are usually recovered from the compound sodium iodate, which is contained in saltpeter, or potassium nitrate, mined in Chile. At one time Chile was among the largest producers of iodine. More recently, large quantities of iodine have been produced from the ashes of burned seaweed. The ocean contains large quantities of iodine concentrated in various kinds of seaweed, such as the giant kelp that grows off the coast of California. Iodine can also be recovered from the natural brines found in underground wells in Arkansas and Oklahoma. Chlorine gas is usually passed through the brine to liberate the iodine.

Although iodine is a poison, one of its most common uses is in the antiseptic solution known as tincture of iodine, which is a 50 percent solution of iodine in alcohol. Although tincture of iodine is no longer the antiseptic of choice for treating minor injuries to the skin, generations of children were subjected to the application of this stinging, dark-brown solution on minor bruises and scratches. Because iodine kills bacteria, campers often use it in the form of tablets or crystals to disinfect water. Iodine salts, such as potassium iodide, are produced in large

Iodine was discovered in 1811 by a scientist who noticed a purple vapor emanating from a sample of kelp to which he had added sulfuric acid. Iodine gets its name from the Greek word iodes, *meaning "violet-colored."*

quantities as additives for table salt and animal feed and constitute an important dietary supplement for people who live inland and do not get enough food from the sea. This is done because iodine is an important constituent of the growth hormone thyroxine, produced in the thyroid gland, and therefore helps to ensure that the thyroid gland functions properly. A deficiency of thyroxine can result in the disease known as goiter. Iodized salt contains approximately 0.01 percent potassium iodide.

Iodine in the form of silver iodide is used in the preparation of photographic film and paper and as a cloud-seeding particle for rainmaking, as discussed in the section on silver. Silver iodide is a preferred agent for seeding clouds because it has the ability to form enormous numbers of tiny crystals. It has been estimated that 1 gram of silver iodide can form as many as 1 million billion seed crystals, which act as nuclei for raindrop formation.

A radioisotope of iodine, iodine-131, with a half-life of 8.1 days, is of great value in diagnosing thyroid diseases. It is usually administered in the form of a solution of radioactive sodium iodide that the patient drinks. Any iodine taken up by the body has a tendency to concentrate in the thyroid gland. Measuring therate at which the gland takes up radioactive sodium iodide after the patient drinks a solution of it tells the physician how well the thyroid is functioning in the manufacture of thyroxine. Its half-life of 8 days is short enough to ensure that most of the

Because iodine kills bacteria, campers often use it in the form of tablets or crystals to disinfect water.

Workers at Eastman Kodak prepare photographic film. Silver iodide is more sensitive to light than silver bromide and is used in the emulsions of high-speed black-and-white film.

iodine-131 in the solution will be eliminated from the body within several weeks.

The ability of iodine to concentrate in the thyroid gland also permits it to be used to treat severe illnesses of the thyroid, such as thyroid cancer. In this case, the radiation emitted by iodine-131 that concentrates in the thyroid gland destroys the cancer cells. This treatment causes minimum damage to other tissue and organs.

Small radioactive iodine pellets, the size of rice grains, are often used as a treatment for prostate cancer. With a half-life of 60 days, the iodine-125 isotope can deliver powerful doses of radiation to destroy the cancer and then disappear after several months. An isotope of palladium with a half-life of 17 days is also used for treatment. Hollow needles are used to carefully implant the tiny pellets into the prostate gland so that there is a minimum of damage to other organs of the body.

A popular chemistry demonstration uses iodine to detect starch. When iodine is applied to a potato, which is rich in starch, it turns a deep blue.

Xenon

IA	IA	IIIB	IVB	VB	VIB	VIIB	VIIIB			IB	IIB	IIIA	IVA	VA	VIA	VIIA	VIIIA
H																	He
Li	Be											B	C	N	O	F	Ne
Na	Mg											Al	Si	P	S	Cl	Ar
K	Ca	Sc	Ti	V	Cr	Mn	Fe	Co	Ni	Cu	Zn	Ga	Ge	As	Se	Br	Kr
Rb	Sr	Y	Zr	Nb	Mo	Tc	Ru	Rh	Pd	Ag	Cd	In	Sn	Sb	Te	I	Xe
Cs	Ba	*La	Hf	Ta	W	Re	Os	Ir	Pt	Au	Hg	Tl	Pb	Bi	Po	At	Rn
Fr	Ra	†Ac	Rf	Db	Sg	Bh	Hs	Mt	Ds	Rg	Cn	Uut	Uuq	Uup	Uuh	Uus	Uuo

*	Ce	Pr	Nd	Pm	Sm	Eu	Gd	Tb	Dy	Ho	Er	Tm	Yb	Lu
†	Th	Pa	U	Np	Pu	Am	Cm	Bk	Cf	Es	Fm	Md	No	Lr

Atomic Number **54**

Chemical Symbol **Xe**

Group **VIIIA—The Noble Gases**

Xenon is a noble gas that exists in the atmosphere in only trace amounts. One of the heaviest of the noble gas family, xenon has a density that is approximately five times that of air. Like the other gases in the family, it exists as a monatomic molecule that has no color, odor, or taste.

It was discovered by the English chemist Sir William Ramsay in 1898 and named by him. He chose the name from the Greek word *xenos*, which means the strange one. Ramsay succeeded in isolating xenon from a sample of air using fractional distillation. Xenon is produced commercially today by essentially the same technique.

Until recently, the noble gases were thought to be completely inert, reacting with no other substances, not even themselves. All of the noble gases have monatomic molecules. In 1962, however, Neil Bartlett, an English chemist working at the University of British Columbia, made the first noble gas compound. He combined xenon and platinum hexafluoride and much to his astonishment obtained a solid, yellow-orange compound. It consisted of molecules made up of xenon, platinum, and fluorine. Bartlett's success inspired other scientists. Within months, the search for other xenon compounds was taken up by scientists at the Argonne National Laboratory near Chicago. They discovered that xenon reacts directly with fluorine at 400°C and that, in the presence of sunlight, it reacts with fluorine even at room temperature. Their work led to three different xenon fluorides: xenon difluoride, xenon tetrafluoride, and xenon hexafluoride. They also managed to synthesize a number of xenon-oxygen

An aerial view of Argonne National Laboratory, located 25 miles outside of Chicago. In the 1960s, scientists there successfully created several compounds of xenon.

compounds. To date, of the noble gases, only xenon and krypton have been shown to form compounds.

Like all of the noble gases, xenon is used in electrical discharge tubes to produce light. When excited, it emits a brilliant white light. Like krypton, it responds rapidly to an electrical current, and a mixture of krypton and xenon is used in high-intensity, short-exposure photographic flash tubes. For this same reason, xenon is also the gas used in rapidly flashing strobe lamps, which supply bursts of intense light for extremely short periods. Cameras use these bursts to "freeze" a rapidly moving object, such as a bullet fired from a gun or a piece of glass in the process of shattering, on photographic film.

A tank filled with more than 750 pounds of ultrapure liquid xenon will play a key role in experiments that will take place during 2011–2015, some 5,000 feet below a mountain in Italy. The experiment, known as Xenon1T, is led by scientists from Columbia University in New York and Rice University in Houston, Texas, in a consortium at the underground Gran Sasso National Laboratory in Italy. Their search is for dark matter, the name given to 25 percent of the universe that theorists say should exist, but has never been seen. The hope is that subatomic particles called weakly interacting massive particles (WIMPs) will be weakly interacting enough to penetrate the mile of earth and rock designed to filter out extraneous radiation, such as cosmic rays, and interact with the xenon. Telltale scintillations of light produced by the interaction of the WIMPs with nuclei of xenon should signal the existence of dark matter.

Cesium

IA																	VIIIA
H	IA											IIIA	IVA	VA	VIA	VIIA	He
Li	Be											B	C	N	O	F	Ne
Na	Mg	IIIB	IVB	VB	VIB	VIIB	VIIIB			IB	IIB	Al	Si	P	S	Cl	Ar
K	Ca	Sc	Ti	V	Cr	Mn	Fe	Co	Ni	Cu	Zn	Ga	Ge	As	Se	Br	Kr
Rb	Sr	Y	Zr	Nb	Mo	Tc	Ru	Rh	Pd	Ag	Cd	In	Sn	Sb	Te	I	Xe
Cs	Ba	*La	Hf	Ta	W	Re	Os	Ir	Pt	Au	Hg	Tl	Pb	Bi	Po	At	Rn
Fr	Ra	†Ac	Rf	Db	Sg	Bh	Hs	Mt	Ds	Rg	Cn	Uut	Uuq	Uup	Uuh	Uus	Uuo

*	Ce	Pr	Nd	Pm	Sm	Eu	Gd	Tb	Dy	Ho	Er	Tm	Yb	Lu
†	Th	Pa	U	Np	Pu	Am	Cm	Bk	Cf	Es	Fm	Md	No	Lr

Atomic Number 55

Chemical Symbol Cs

Group IA—The Alkali Metals

Cs

Cesium is an alkali metal, a member of the group of metals that occupy the first column in the periodic table. Like all of the alkali metals, it is soft and silvery in its pure state and is, in fact, the softest metal known. It is also one of the most reactive of the alkali metals, reacting violently with water to release hydrogen gas. Like the alkali metals potassium and rubidium, cesium reacts with oxygen to produce superoxides of the general formula CsO_2. When they react with water or carbon dioxide, these superoxides release oxygen. This property makes them very useful in self-contained breathing apparatuses used by emergency workers and fire fighters in regions where toxic fumes are released.

Cesium has the extremely low melting point of 29°C. Since the human body temperature is 37°C, cesium will melt when held in the hand. The only metal with a lower melting point is mercury.

Cesium was discovered in 1860 by the German chemists Robert Bunsen and Gustav Kirchhoff. They isolated it by observing the colors, or spectral emission lines, that cesium emits when it is heated. The observed lines were bright blue, which prompted the discoverers to name the element from the Latin *coesius*, which means "sky blue."

Cesium was discovered in 1860 by the German chemists Gustav Kirchhoff (above) and Robert Bunsen.

Cesium is far too active to be found in its free metallic state. Its major source is the mineral pollucite, a compound of silicon, aluminum, cesium, and oxygen.

The extreme reactivity of cesium has made it useful in removing unwanted gases from vacuum systems. It is important, for example, in ensuring that the elements inside a television tube are kept in a vacuum, free from potentially harmful extraneous gases. A small amount of cesium in the interior of the tube acts as a scavenger by immediately reacting with such unwanted gas.

In 1960, an International Committee of Weights and Measures chose cesium-133, the only naturally occurring isotope of cesium, as the world's official measure of time. The second is now measured in terms of the radiation emitted by a cesium-133 atom when it is excited by an external energy source rather than in terms of the Earth's rotation around the sun, as it used to be. The second is now officially defined as the duration (the elapsed time) of exactly 9,192,631,770 vibrations of the radiation emitted by such an atom. Cesium iodide crystals are often used in a radiation detector known as a scintillation detector. Radiation interacting with the crystals produces scintillations of light, which are then converted into electrical signals.

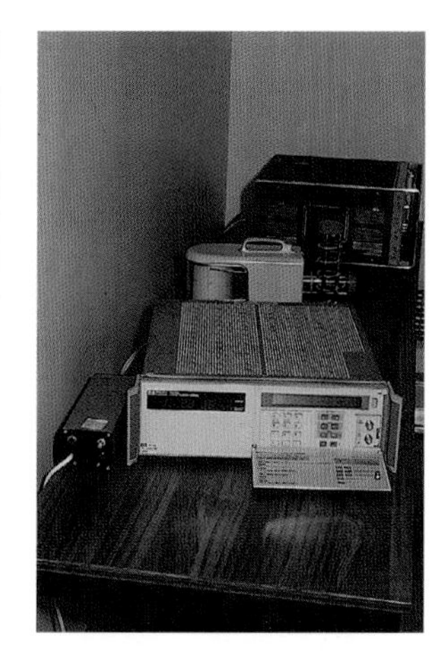

Atomic clocks, like the one shown here, contain an isotope of cesium. They are used around the world to establish an official standard of time.

Liquid metal cesium is being used as a propellant in the European Space Agency's (ESA) design for the smallest engine ever used in space. It is known as the Field Emission Electric Propulsion (FEEP) engine and has an astonishing width of only 4 inches. It uses an electric field to accelerate electrically charged cesium atoms, called ions, through an extremely thin slit to produce thrust.

It is planned to operate aboard the Laser Interferometer Space Antenna (LISA) Pathfinder, a joint NASA–ESA mission to detect the gravitational waves that Albert Einstein predicted in his general theory of relativity. The Pathfinder is scheduled to be launched in late 2011 and will orbit the Earth at a distance of 1.5 million kilometers, known as Lagrange Point L1, where the gravitational fields of the Earth and the Sun cancel one another. Although the thrust produced by the FEEP will be extremely small, it will have the advantage of being very precisely controllable, sensitive, and accurate enough to counterbalance the force of sunlight that LISA will be subjected to.

Barium

IA																	VIIIA
H	IA											IIIA	IVA	VA	VIA	VIIA	He
Li	Be											B	C	N	O	F	Ne
Na	Mg	IIIB	IVB	VB	VIB	VIIB	VIIIB			IB	IIB	Al	Si	P	S	Cl	Ar
K	Ca	Sc	Ti	V	Cr	Mn	Fe	Co	Ni	Cu	Zn	Ga	Ge	As	Se	Br	Kr
Rb	Sr	Y	Zr	Nb	Mo	Tc	Ru	Rh	Pd	Ag	Cd	In	Sn	Sb	Te	I	Xe
Cs	Ba	*La	Hf	Ta	W	Re	Os	Ir	Pt	Au	Hg	Tl	Pb	Bi	Po	At	Rn
Fr	Ra	†Ac	Rf	Db	Sg	Bh	Hs	Mt	Ds	Rg	Cn	Uut	Uuq	Uup	Uuh	Uus	Uuo

*	Ce	Pr	Nd	Pm	Sm	Eu	Gd	Tb	Dy	Ho	Er	Tm	Yb	Lu
†	Th	Pa	U	Np	Pu	Am	Cm	Bk	Cf	Es	Fm	Md	No	Lr

Ba

Barium is a soft, silvery-white metal that burns easily in air and reacts with water to produce hydrogen. As an alkaline-earth metal, it is extremely reactive. Barium is a fairly abundant element, ranking sixth in abundance among the elements that make up the Earth's crust. It is found in the minerals witherite, a barium carbonate ($BaCO_3$), and baryte, a barium sulfate ($BaSO_4$). The element was discovered by the English chemist Sir Humphry Davy in 1808.

Barium is quite toxic. If ingested in the form of a soluble salt, such as barium chloride, it seriously impairs the heart's functioning and can produce the condition known as ventricular fibrillation, in which the heart beats erratically. On the other hand, barium in an insoluble form, such as in barium sulfate, can be swallowed without any damage to the body. Indeed, radiologists use barium sulfate to examine a patient's intestinal tract with X rays. This procedure is possible because barium sulfate is a highly dense salt that is opaque to X rays. When a patient either drinks a suspension of barium sulfate in water or is given a barium enema, consisting of a similar suspension, the barium sulfate particles temporarily fill the digestive tract. Since the barium sulfate absorbs X rays and blocks their transmission, the intestines and other organs will appear white in an X-ray picture, clearly showing the shape of the intestines and other organs outlined against a black background.

Barium metal has few commercial applications because of its readiness to react with oxygen and moisture. This property does make it useful as a scavenger in vacuum systems by removing unwanted gases. As a constituent of an alloy with other

Atomic Number **56**

Chemical Symbol **Ba**

Group **IIA—The Alkaline-Earth Metals**

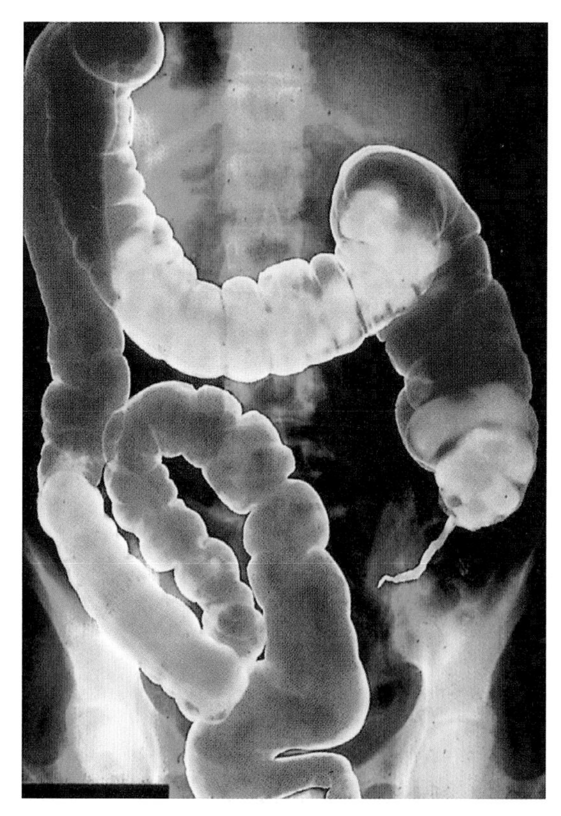

After a patient ingests a solution of barium sulfate, a doctor can take an X ray of the person's digestive tract, which appears white wherever the barium sulfate accumulates.

Barium was discovered by the English chemist Sir Humphry Davy in 1808.

metals, however, it is used to make spark plugs, because it emits electrons very easily when heated.

Barium sulfate also has a number of uses based on its low solubility in water and brilliant white color. It is used as a whitener in photographic papers and as a filler in writing paper, plastics, and artificial fibers. A mixture of zinc sulfide and barium sulfate is used as a bright white paint pigment called lithopone. Paradoxically, the green color of fireworks is often produced by barium compounds.

Lanthanum

Atomic Number **57**

Chemical Symbol **La**

Group IIIB—Rare Earth Element (Lanthanides)

IA	IA	IIIB	IVB	VB	VIB	VIIB	VIIIB			IB	IIB	IIIA	IVA	VA	VIA	VIIA	VIIIA
H																	He
Li	Be											B	C	N	O	F	Ne
Na	Mg											Al	Si	P	S	Cl	Ar
K	Ca	Sc	Ti	V	Cr	Mn	Fe	Co	Ni	Cu	Zn	Ga	Ge	As	Se	Br	Kr
Rb	Sr	Y	Zr	Nb	Mo	Tc	Ru	Rh	Pd	Ag	Cd	In	Sn	Sb	Te	I	Xe
Cs	Ba	*La	Hf	Ta	W	Re	Os	Ir	Pt	Au	Hg	Tl	Pb	Bi	Po	At	Rn
Fr	Ra	†Ac	Rf	Db	Sg	Bh	Hs	Mt	Ds	Rg	Cn	Uut	Uuq	Uup	Uuh	Uus	Uuo

*	Ce	Pr	Nd	Pm	Sm	Eu	Gd	Tb	Dy	Ho	Er	Tm	Yb	Lu
†	Th	Pa	U	Np	Pu	Am	Cm	Bk	Cf	Es	Fm	Md	No	Lr

La

Lanthanum is the first element of a special series of elements called the rare earths. This series, also called the lanthanide series, consists of lanthanum and the next 14 elements that follow it across the periodic table. The last member of the series is lutetium (atomic number 71). All of the elements that make up this series are characterized by having their added electrons, the number of which increases with the atomic number of each element in the series, buried in the interior of the atoms of each element rather than contained in the outermost, or valence, orbital. Since the outermost electrons have the greatest influence on the properties of the elements, it should not be surprising that the variations in the properties of the rare earth elements are not as dramatic as the variations observed as one moves across the group-A elements, to take but one example. Many of the properties of lanthanum are, for instance, more or less typical of all the elements in the rare earth series. It is quite common, for example, to find most of the rare earth elements, usually in the form of their oxide compounds, mixed together in the minerals that contain these elements.

It is this mixing that has led to the name "rare." The abundance of the rare earths in the Earth's crust is actually comparable to that of such common metals as zinc, tin, and lead. But unlike these metals, a rare earth element is not concentrated in a geological vein or outcropping that can be easily mined.

Lanthanum was discovered in 1839 by the Swedish chemist Carl Gustaf Mosander, who extracted it as an impurity from a rare earth mineral called cerium nitrate. It was named from the Greek word *lanthanein*, meaning "to lie hidden."

Lanthanum was discovered in 1839 by a Swedish chemist who extracted it from a mineral called cerium nitrate. It was named from the Greek word lanthanein, *meaning "to lie hidden."*

Lanthanum is a silvery-white metal, quite reactive, ductile and malleable, and soft enough to be cut with a knife. It forms an oxide fairly rapidly in air and reacts vigorously in warm water, liberating hydrogen gas. China is today the dominant producer of the rare earths. Its mines in Inner Mongolia supply some 97 percent of the world's requirements. Smaller amounts are found in bastnasite ores from the Mojave Desert, California, and monazite-bearing river sands located chiefly in India, Brazil, and South Africa. Lanthanum can be recovered from the monazite, which contains all the rare earths as well as thorium and calcium, by ion exchange (see praseodymium).

Interest in the rare earth elements has risen dramatically as many of them have become essential for a host of modern consumer products such as flat-screen televisions, touch screens, iPhones, cell phones, and wind turbines, to name just a few.

Lanthanum, for example, plays an essential role in the NiMH (nickel–metal hydride) battery used in electric and hybrid automobiles. The Prius, the hybrid manufactured by Toyota, contains about 25 pounds of lanthanum in its NiMH battery.

Another commercial use of lanthanide compounds is in fabricating the electrodes for the high-intensity carbon arc lamps used in searchlights, studio lighting, and motion-picture projectors. Lanthanum oxide improves the resistance of glass to attack by alkaline substances and is used in making special glass for sophisticated optical equipment. Lanthanum and its compounds are considered moderately toxic.

Lanthanum and its isotopes are found in the fission fragments that are produced when uranium fissions. During the fission process the uranium nucleus, after absorbing a neutron, usually "splits" into two fragments whose nuclei are of intermediate size. There are many different ways of splitting, so that a great many mid-size elements are created. Lanthanum isotopes are among those formed. It was the discovery of these isotopes, as well as those of barium, by the German chemist Otto Hahn, that eventually led to the idea of nuclear fission. He published his findings in January 1939 and ended the report by stating that "it violated all previous experience in the field of nuclear physics."

It was the German physicist Lise Meitner who first understood the significance of this discovery, and together with her nephew, the German physicist Otto Frisch, correctly guessed that when a uranium nucleus absorbs a neutron, it becomes unstable and divides into two parts, like the reproductive fissioning of a living cell.

Cerium

IA	IA	IIIB	IVB	VB	VIB	VIIB	VIIIB			IB	IIB	IIIA	IVA	VA	VIA	VIIA	VIIIA
H																	He
Li	Be											B	C	N	O	F	Ne
Na	Mg											Al	Si	P	S	Cl	Ar
K	Ca	Sc	Ti	V	Cr	Mn	Fe	Co	Ni	Cu	Zn	Ga	Ge	As	Se	Br	Kr
Rb	Sr	Y	Zr	Nb	Mo	Tc	Ru	Rh	Pd	Ag	Cd	In	Sn	Sb	Te	I	Xe
Cs	Ba	*La	Hf	Ta	W	Re	Os	Ir	Pt	Au	Hg	Tl	Pb	Bi	Po	At	Rn
Fr	Ra	†Ac	Rf	Db	Sg	Bh	Hs	Mt	Ds	Rg	Cn	Uut	Uuq	Uup	Uuh	Uus	Uuo

*	Ce	Pr	Nd	Pm	Sm	Eu	Gd	Tb	Dy	Ho	Er	Tm	Yb	Lu
†	Th	Pa	U	Np	Pu	Am	Cm	Bk	Cf	Es	Fm	Md	No	Lr

Ce

Cerium is the most abundant of the metals that make up the rare earth elements. It was simultaneously discovered by the Swedish chemist Jöns Jakob Berzelius and the German chemists Wilhelm Hisinger and Martin Klaproth in 1803, as an impurity in the mineral bastnasite. It was named for the asteroid Ceres, whose discovery in 1801 had excited the scientific world.

China is today the dominant producer of the rare earth elements. Its mines in Inner Mongolia supply some 97 percent of the world's requirements. Smaller amounts of cerium are obtained from river sands containing monazite and from bastnasite ore (see lanthanum).

Cerium in its pure metallic form was not prepared until 1875. It is an iron-gray metal that is quite malleable and ductile. It is also one of the most reactive of the rare earth metals, oxidizing quite readily in moist air and decomposing fairly rapidly in heated water. The metal has been known to ignite spontaneously in air from the heat generated by scratching it.

Cerium compounds, like those of lanthanum, are used commercially to form the electrodes of the high-intensity carbon arc lamps used in searchlights and motion-picture projectors. As an oxide, cerium is used as an additive to the walls of "self-cleaning" ovens, where it seems to prevent the buildup of cooking residues. Its oxide is also used to polish lenses for cameras and telescopes.

Atomic Number **58**

Chemical Symbol **Ce**

Group **IIIB—Rare Earth Element (Lanthanides)**

The walls of some self-cleaning ovens are lined with a cerium compound that prevents the buildup of cooking residues.

IA																	VIIIA
H	IA											IIIA	IVA	VA	VIA	VIIA	He
Li	Be											B	C	N	O	F	Ne
Na	Mg	IIIB	IVB	VB	VIB	VIIB		VIIIB		IB	IIB	Al	Si	P	S	Cl	Ar
K	Ca	Sc	Ti	V	Cr	Mn	Fe	Co	Ni	Cu	Zn	Ga	Ge	As	Se	Br	Kr
Rb	Sr	Y	Zr	Nb	Mo	Tc	Ru	Rh	Pd	Ag	Cd	In	Sn	Sb	Te	I	Xe
Cs	Ba	*La	Hf	Ta	W	Re	Os	Ir	Pt	Au	Hg	Tl	Pb	Bi	Po	At	Rn
Fr	Ra	†Ac	Rf	Db	Sg	Bh	Hs	Mt	Ds	Rg	Cn	Uut	Uuq	Uup	Uuh	Uus	Uuo

*	Ce	Pr	Nd	Pm	Sm	Eu	Gd	Tb	Dy	Ho	Er	Tm	Yb	Lu
†	Th	Pa	U	Np	Pu	Am	Cm	Bk	Cf	Es	Fm	Md	No	Lr

Praseodymium

Atomic Number 59

Chemical Symbol Pr

Group IIIB—Rare Earth Element (Lanthanides)

Pr

Praseodymium, pronounced pra-si-oh-DYM-ium, is a rare earth metal that was finally isolated and identified as an element in 1885 by Carl Auer von Welsbach, an Austrian baron who had a love of mineralogy. He separated a mineral called didymium into two new distinct elemental salts, one of which contained praseodymium. It was named from two Greek words—*prasios*, meaning "green," and *didymos*, meaning "twin." This name refers to the green color of the element's oxide. The other salt contained the element neodymium.

China is today the dominant producer of the rare earths, with 97 percent of the world's requirements coming from its mines in Inner Mongolia. Smaller quantities are found in monazite and bastnasite ores. Monazite contains all of the rare elements, often in concentrations as high as 50 percent. It is found in river sand in India, Brazil, and South Africa. Bastnasite deposits are found in the Mojave Desert in California, but for economic reasons have been mined only in small quantities. Praseodymium metal is usually isolated from its ore by ion-exchange techniques. As the name implies, an ion-exchange process is used to isolate one kind of ion by substituting it with another. In one type of exchange process, the active ingredient is usually a resin made up of large molecules that have a netlike structure. The resin contains very mobile ions, such as sodium or potassium ions, that are loosely connected to the net. When a solution containing other ions is passed through the netlike structure of the resin, they replace the sodium or potassium ions that then diffuse out of the net.

Another type of resin consists of tiny plastic beads that have certain ions loosely attached to their surface. When a solution of

Praseodymium was not prepared in its pure metallic form until 1931.

the ion to be removed is passed through a column of these beads, the ions in solution replace the existing ions and are separated from the solution.

Praseodymium was not prepared in its pure metallic form until 1931. It is a soft, silvery-white metal that is malleable and ductile. Because it tarnishes to form a green oxide in air, it is usually kept sealed in plastic or covered with oil.

As with many of the rare earths, the oxide of praseodymium is used to fabricate the electrodes of the high-intensity carbon arc lamps used in searchlights and motion-picture projectors. Small quantities of the metal are added to magnesium to form an alloy that is both stronger and more corrosion-resistant than the original metals. This alloy is used to make both automobile and aircraft parts. Another alloy of interest, called Mischmetal, contains about 5 percent praseodymium and uses the tendency of praseodymium to give off sparks when it is scratched. Mischmetal uses this feature to make flints for cigarette lighters.

Didymium glass, named from the Greek for "twin elements," is made from a mixture of praseodymium and neodymium and is used in the safety goggles worn by glassblowers and welders. The glass absorbs the bright yellow–orange flare emitted by the sodium in the heated glass, so that it is possible to safely see through the glow as the glass or metal is being worked on.

IA	IA	IIIB	IVB	VB	VIB	VIIB	VIIIB			IB	IIB	IIIA	IVA	VA	VIA	VIIA	VIIIA
H																	He
Li	Be											B	C	N	O	F	Ne
Na	Mg											Al	Si	P	S	Cl	Ar
K	Ca	Sc	Ti	V	Cr	Mn	Fe	Co	Ni	Cu	Zn	Ga	Ge	As	Se	Br	Kr
Rb	Sr	Y	Zr	Nb	Mo	Tc	Ru	Rh	Pd	Ag	Cd	In	Sn	Sb	Te	I	Xe
Cs	Ba	*La	Hf	Ta	W	Re	Os	Ir	Pt	Au	Hg	Tl	Pb	Bi	Po	At	Rn
Fr	Ra	†Ac	Rf	Db	Sg	Bh	Hs	Mt	Ds	Rg	Cn	Uut	Uuq	Uup	Uuh	Uus	Uuo

*	Ce	Pr	Nd	Pm	Sm	Eu	Gd	Tb	Dy	Ho	Er	Tm	Yb	Lu
†	Th	Pa	U	Np	Pu	Am	Cm	Bk	Cf	Es	Fm	Md	No	Lr

Neodymium

Atomic Number **60**

Chemical Symbol **Nd**

Group IIIB—Rare Earth Element (Lanthanides)

Nd

Neodymium is a rare earth metal discovered in 1885 by an Austrian amateur mineralogist, Baron Carl Auer von Welsbach. He separated a mineral called didymium into two new distinct elemental salts, one of which contained neodymium. Welsbach created the name for this element from two Greek words, *neos*, meaning "new," and *didymos*, meaning "twin." The other salt contained the element praseodymium.

Small quantities of neodymium are found in monazite ores found in river sand in India, Brazil, and South Africa and bastnasite ores from the Mojave Desert in California. But the mines of Inner Mongolia, in China, are by far the dominant producers of the rare earths today. The neodymium is isolated from its ores by ion exchange (see praseodymium).

Pure neodymium metal was not isolated until 1925. It has a silver luster that quickly tarnishes in air and is therefore usually stored in a plastic wrapping or covered with mineral oil.

Neodymium and its oxides, which tend to be rose colored, are used to make colored glass for special purposes. Some examples of such glass are the colored glass in welder's goggles and the "artificial ruby" used as a substitute for real rubies in certain types of lasers. Mischmetal, an important compound used for making steel alloys, contains about 18 percent neodymium.

About 24 percent of naturally occurring neodymium consists of the very weakly radioactive isotope neodymium-144. Because this isotope has a half-life of 2 million billion years, very little disintegrates during a short period.

Neodymium is a magnetic substance and is used to create some of the most powerful magnets on Earth. The neodymium

A neodymium disc magnet with a diameter of only half an inch is strong enough to respond to magnetic materials in the printing ink used for paper money, and the element can be used to detect counterfeit currency.

supermagnets, known as NIB magnets since they consist of *ne*odymium, *i*ron, and *b*oron, are so strong that two small magnets will press firmly to either side of one's hand without falling. Neodymium magnets will cling to each other with such force that it is almost impossible for most people to separate them by trying to pull them apart. Students and researchers are usually cautioned to wear eye goggles when working with these magnets, since they might inadvertently clamp together with such force that they fracture. Small splinters and fragments flying off from the point of contact might cause injuries to the eyes.

What makes the neodymium magnets especially useful is that they are relatively inexpensive. A small but powerful half-inch disc costs approximately 10 dollars. These disc magnets are strong enough to respond to magnetic materials in the printing ink used for paper money and can be used to detect counterfeit currency.

Supermagnets made of neodymium have become an indispensable component of the wind turbines that are being developed to supply a renewable source of electrical energy. With their low weight and strong magnetic field, neodymium magnets are ideal for an electricity-producing system powered by a few slowly spinning blades sitting on top of tall towers. It is estimated to require approximately 1 ton of neodymion per megawatt of generating capacity.

IA																	VIIIA
H	IA											IIIA	IVA	VA	VIA	VIIA	He
Li	Be											B	C	N	O	F	Ne
Na	Mg	IIIB	IVB	VB	VIB	VIIB	VIIIB			IB	IIB	Al	Si	P	S	Cl	Ar
K	Ca	Sc	Ti	V	Cr	Mn	Fe	Co	Ni	Cu	Zn	Ga	Ge	As	Se	Br	Kr
Rb	Sr	Y	Zr	Nb	Mo	Tc	Ru	Rh	Pd	Ag	Cd	In	Sn	Sb	Te	I	Xe
Cs	Ba	*La	Hf	Ta	W	Re	Os	Ir	Pt	Au	Hg	Tl	Pb	Bi	Po	At	Rn
Fr	Ra	†Ac	Rf	Db	Sg	Bh	Hs	Mt	Ds	Rg	Cn	Uut	Uuq	Uup	Uuh	Uus	Uuo

*	Ce	Pr	Nd	Pm	Sm	Eu	Gd	Tb	Dy	Ho	Er	Tm	Yb	Lu
†	Th	Pa	U	Np	Pu	Am	Cm	Bk	Cf	Es	Fm	Md	No	Lr

Promethium is a synthetic rare earth element made in nuclear accelerators and nuclear reactors. No trace of the element has been found in the Earth's crust. It is named for the Greek god Prometheus who stole fire from the heavens. As befits its name, promethium has been identified in the spectrum of several stars in the Andromeda Galaxy.

Promethium was first discovered by a team of scientists at Oak Ridge National Laboratory in 1947. The laboratory continues to conduct research on the element and its properties. Here, promethium oxide is loaded into a capsule for testing.

No trace of promethium has been found in the Earth's crust. Promethium has been identified in the spectrum of several stars in the Andromeda Galaxy.

Study of the periodic table led several scientists at the beginning of the 20th century to predict the existence of an element between neodymium and samarium. The element whose atomic number was 61 was simply missing. Many nuclear facilities throughout the United States tried to create and identify the element. Its existence was finally established in 1947 by J. A. Mirinsky, L. E. Glendenin, and C. D. Coryell, a team of scientists working at the National Laboratory at Oak Ridge, Tennessee. They found promethium among the many different nuclei of intermediate size, known as fission fragments, produced by the splitting of uranium atoms in a nuclear reactor. When neodymium is subjected to the intense neutron radiation present in a reactor, it is converted into promethium.

Twenty-eight isotopes of promethium have so far been synthesized, all of which are radioactive. Promethium-145 is the longest-lived of these isotopes with a half-life of 17.7 years, although promethium-147, an electron-emitting isotope with a half-life of 2.6 years, is the most useful for scientific applications. The electrons it emits can produce light from certain phosphors, and the light is then used in self-luminous products. Using photocells to convert this phosphor-generated light into electricity can also produce nuclear-powered batteries. These batteries have a useful life of about 5 years and are generally designed for spacecraft and weapons guidance systems. The electrons can also be used in the production of portable X-ray machines and as a thickness gauge for measuring the thickness of paper and sheet metal as it rolls by on an industrial assembly line.

Very little is known of the chemical and physical properties of pure promethium. However, promethium salts have been made, and the radiation they emit causes the surrounding air to glow in the dark with a blue light.

IA																	VIIIA
H	IA											IIIA	IVA	VA	VIA	VIIA	He
Li	Be											B	C	N	O	F	Ne
Na	Mg	IIIB	IVB	VB	VIB	VIIB	VIIIB			IB	IIB	Al	Si	P	S	Cl	Ar
K	Ca	Sc	Ti	V	Cr	Mn	Fe	Co	Ni	Cu	Zn	Ga	Ge	As	Se	Br	Kr
Rb	Sr	Y	Zr	Nb	Mo	Tc	Ru	Rh	Pd	Ag	Cd	In	Sn	Sb	Te	I	Xe
Cs	Ba	*La	Hf	Ta	W	Re	Os	Ir	Pt	Au	Hg	Tl	Pb	Bi	Po	At	Rn
Fr	Ra	†Ac	Rf	Db	Sg	Bh	Hs	Mt	Ds	Rg	Cn	Uut	Uuq	Uup	Uuh	Uus	Uuo

*	Ce	Pr	Nd	Pm	Sm	Eu	Gd	Tb	Dy	Ho	Er	Tm	Yb	Lu
†	Th	Pa	U	Np	Pu	Am	Cm	Bk	Cf	Es	Fm	Md	No	Lr

Samarium

Samarium is a rare earth metal that is fifth in abundance among the rare earth elements in the Earth's crust. It was discovered in 1879 by the French chemist Paul-Émile Lecoq de Boisbaudran, who correctly identified the element's spectral absorption lines in a mineral called samarskite, which had been named for a Russian army officer, Colonel Samarski, a distinguished engineer. De Boisbaudran decided to name his newly discovered element for the mineral.

As is the case with all of the rare earths, China is the dominant producer, with its mines in Inner Mongolia supplying some 97 percent of the world's requirements. Smaller quantities of samarium are found in monazite and bastnasite ores. Monazite contains all of the rare elements, often in concentrations as high as 50 percent. It is found in river sand in India, Brazil, and South Africa. Bastnasite deposits are found in the Mojave Desert in California, but for economic reasons are only being mined in small quantities. The samarium is isolated from its ores by ion exchange (see praseodymium).

In its pure metallic form, samarium has a silvery-white luster and is fairly resistant to oxidation in air. The metal will, however, ignite spontaneously at temperatures as low as 150°C. Permanent magnets made of an alloy of samarium and cobalt (SmCo) are second only to neodymion magnets in their strength. They are also among the most difficult to demagnetize, which makes them useful for such devices as headphones and electric guitar pickups. Samarium oxide is an excellent absorber of infrared radiation and is added for this purpose to special types of glass and infrared-sensitive phosphors.

One of the radioactive isotopes of samarium, samarium-153, also called Quadramet, is taken up by bone tissue and is used in cases of bone cancer to alleviate the pain associated with the disease.

Atomic Number **62**

Chemical Symbol **Sm**

Group **IIIB—Rare Earth Element (Lanthanides)**

Europium

Atomic Number 63

Chemical Symbol Eu

Group IIIB—Rare Earth Element (Lanthanides)

IA	IA	IIIB	IVB	VB	VIB	VIIB	VIIIB			IB	IIB	IIIA	IVA	VA	VIA	VIIA	VIIIA
H																	He
Li	Be											B	C	N	O	F	Ne
Na	Mg											Al	Si	P	S	Cl	Ar
K	Ca	Sc	Ti	V	Cr	Mn	Fe	Co	Ni	Cu	Zn	Ga	Ge	As	Se	Br	Kr
Rb	Sr	Y	Zr	Nb	Mo	Tc	Ru	Rh	Pd	Ag	Cd	In	Sn	Sb	Te	I	Xe
Cs	Ba	*La	Hf	Ta	W	Re	Os	Ir	Pt	Au	Hg	Tl	Pb	Bi	Po	At	Rn
Fr	Ra	†Ac	Rf	Db	Sg	Bh	Hs	Mt	Ds	Rg	Cn	Uut	Uuq	Uup	Uuh	Uus	Uuo

*	Ce	Pr	Nd	Pm	Sm	Eu	Gd	Tb	Dy	Ho	Er	Tm	Yb	Lu
†	Th	Pa	U	Np	Pu	Am	Cm	Bk	Cf	Es	Fm	Md	No	Lr

Eu

Europium is one of the rarest of the rare earth elements. In 1901, the French chemist Eugène-Anatole Demarcay finally isolated an impurity in a samarium–gadolinium sample he was studying and identified the impurity as a new element. He named his new element for the continent of Europe. It took many more years before the pure metal was isolated.

Most of the world's europium comes from the mines in Inner Mongolia. Smaller quantities are found in monazite and bastnasite ores. Monazite contains all of the rare elements, often in concentrations as high as 50 percent. It is found in river sand in India, Brazil, and South Africa. Bastnasite deposits are found in the Mojave Desert in California, but for economic reasons are only being mined in small quantities. The europium is isolated from its ores by ion exchange (see praseodymium). Some europium is also recovered from the fission fragments produced in nuclear reactors.

Pure europium metal is a fairly soft, silvery-white metal. It is quite ductile and is one of the most reactive of the rare earth metals. It oxidizes fairly rapidly in air and reacts with water to produce hydrogen. It will ignite spontaneously in air at temperatures above 150°C.

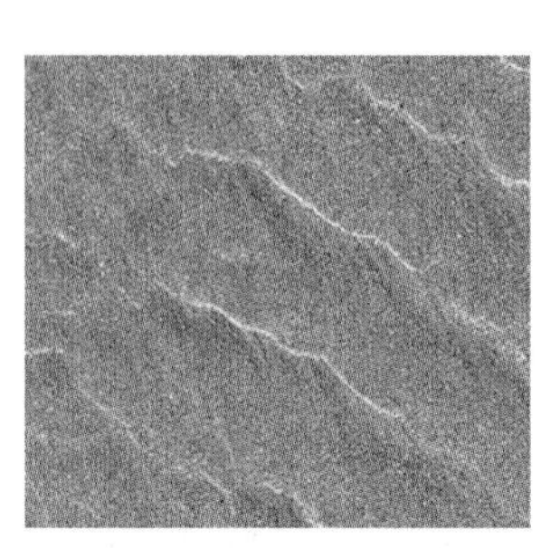

The mineral monazite, found in Florida beach sand, contains traces of all of the rare earth elements—including europium.

Europium oxide is fairly widely used as an additive to improve the efficiency of the red phosphor in color television tubes and computer monitors. It is also used to improve the energy efficiency of fluorescent lamps.

IA	IA	IIIB	IVB	VB	VIB	VIIB	VIIIB			IB	IIB	IIIA	IVA	VA	VIA	VIIA	VIIIA
H																	He
Li	Be											B	C	N	O	F	Ne
Na	Mg											Al	Si	P	S	Cl	Ar
K	Ca	Sc	Ti	V	Cr	Mn	Fe	Co	Ni	Cu	Zn	Ga	Ge	As	Se	Br	Kr
Rb	Sr	Y	Zr	Nb	Mo	Tc	Ru	Rh	Pd	Ag	Cd	In	Sn	Sb	Te	I	Xe
Cs	Ba	*La	Hf	Ta	W	Re	Os	Ir	Pt	Au	Hg	Tl	Pb	Bi	Po	At	Rn
Fr	Ra	†Ac	Rf	Db	Sg	Bh	Hs	Mt	Ds	Rg	Cn	Uut	Uuq	Uup	Uuh	Uus	Uuo

*	Ce	Pr	Nd	Pm	Sm	Eu	Gd	Tb	Dy	Ho	Er	Tm	Yb	Lu
†	Th	Pa	U	Np	Pu	Am	Cm	Bk	Cf	Es	Fm	Md	No	Lr

Gadolinium

Gd

Gadolinium is a rare earth element that is the sixth most common of these elements in the Earth's crust. The honor for its discovery is shared by the French chemists Jean de Marignac and Paul-Émile Lecoq de Boisbaudran. In 1886, de Boisbaudran finally isolated and identified the element from the mineral gadolinite, for which he named it. The mineral itself was named for the distinguished Finnish chemist Johan Gadolin.

Most of the world's supply of gadolinium now comes from China's Inner Mongolian mines. Smaller quantities are found in monazite and bastnasite ores. Monazite contains all of the rare elements, often in concentrations as high as 50 percent, and is found in river sand in India, Brazil, and South Africa. Bastnasite deposits are found in the Mojave Desert in California, but for economic reasons are only being mined in small quantities. The gadolinium is isolated from its ores by ion exchange (see praseodymium).

In its pure metallic form, gadolinium is a shiny, silvery-white metal that is malleable and ductile. It tarnishes very slowly in moist air, forming an oxide that flakes off the surface of the metal. This process exposes fresh gadolinium to the oxygen in the air, and the corrosion continues.

Two isotopes of gadolinium are among the most potent known absorbers of neutrons. Although their scarcity limits their use, they are often used in making the control rods for nuclear reactors, whose function is to absorb neutrons and in so doing effectively stop the chain reaction in a reactor. Gadolinium is also used as an alloying agent in the production of special steels.

Atomic Number **64**

Chemical Symbol **Gd**

Group IIIB—Rare Earth Element (Lanthanides)

Gadolinium was discovered in 1886 and named for the distinguished Finnish chemist Johan Gadolin.

Compounds of gadolinium, like those of many of the rare earth elements, are used to make phosphors for color television tubes and computer monitors. Gadolinium is ferromagnetic, which means that like iron and cobalt, it is strongly attracted by magnets. An interesting property of gadolinium is that its Curie point, the temperature at which a magnetic material loses its magnetism, is only about 17°C, or approximately room temperature.

Gadolinium has proven of value in a new technique for probing the interior of metals, known as neutron radiography. It is used in the airline and shipbuilding industries to search for hidden flaws and structural weaknesses in hulls and fuselages. Like the more traditional X-ray picture, it produces a shadowgram of the materials that make up a structure, since some materials are more opaque to neutrons than others.

In addition to the usual film required for X-rays, an additional screen called a conversion screen is needed in neutron radiography. This screen, commonly made of gadolinium, is placed in close contact with the film. Neutrons that penetrate the sample being investigated interact with the screen, making it radioactive. The radiation then given off by the screen darkens the film and an image is formed.

Gadolinium dye is used as a contrasting agent in the medical procedure known as magnetic resonance imaging. Although it greatly enhances the images of certain diseased and abnormal tissues in the body, there has been some controversy about possible health problems associated with its use.

IA																	VIIIA
H	IA											IIIA	IVA	VA	VIA	VIIA	He
Li	Be											B	C	N	O	F	Ne
Na	Mg	IIIB	IVB	VB	VIB	VIIB	VIIIB			IB	IIB	Al	Si	P	S	Cl	Ar
K	Ca	Sc	Ti	V	Cr	Mn	Fe	Co	Ni	Cu	Zn	Ga	Ge	As	Se	Br	Kr
Rb	Sr	Y	Zr	Nb	Mo	Tc	Ru	Rh	Pd	Ag	Cd	In	Sn	Sb	Te	I	Xe
Cs	Ba	*La	Hf	Ta	W	Re	Os	Ir	Pt	Au	Hg	Tl	Pb	Bi	Po	At	Rn
Fr	Ra	†Ac	Rf	Db	Sg	Bh	Hs	Mt	Ds	Rg	Cn	Uut	Uuq	Uup	Uuh	Uus	Uuo

*	Ce	Pr	Nd	Pm	Sm	Eu	Gd	Tb	Dy	Ho	Er	Tm	Yb	Lu
†	Th	Pa	U	Np	Pu	Am	Cm	Bk	Cf	Es	Fm	Md	No	Lr

Terbium

Tb

Terbium is one of the least abundant of the rare earth elements. Carl Gustaf Mosander discovered the element in 1843 by isolating its oxide from the mineral yttria. He named it for the mineral, which is itself named for Ytterby, a Swedish village where yttria and terbium are often found.

The major source of terbium is from the large rare earth mines in Inner Mongolia, China. China today has assumed a dominant role in supplying rare earths and is responsible for producing almost 97 percent of the world's supply. Small amounts are also found in river sands that contain the mineral monazite. Located chiefly in Brazil, India, and South Africa, monazite is a mixture of rare earth elements combined with thorium and calcium. The terbium is extracted from the sands by ion-exchange techniques (see praseodymium).

In its pure metallic form, terbium is silvery-white, malleable, ductile, and soft enough to be cut with a knife. It bears a resemblance to lead, but it is much heavier. Like lead, it is fairly resistant to corrosion.

One of the major uses of terbium is as a critical component of the light-producing phosphor used in compact fluorescent light bulbs (CFLs). The highly energy efficient CFL, with its familiar helix-like twists of tubing, has been mandated by Congress to completely replace the traditional incandescent light bulb in the United States by 2014.

Compounds of terbium have also found use in special lasers and as phosphors that produce the green color in television tubes and computer monitors. Other applications include the production of alloys with special magnetic properties for use in compact discs and in the fabrication of high-definition X-ray screens.

Atomic Number **65**

Chemical Symbol **Tb**

Group **IIIB—Rare Earth Element (Lanthanides)**

Dysprosium

Atomic Number **66**

Chemical Symbol **Dy**

Group **IIIB—Rare Earth Element (Lanthanides)**

IA	IA	IIIB	IVB	VB	VIB	VIIB	VIIIB			IB	IIB	IIIA	IVA	VA	VIA	VIIA	VIIIA
H																	He
Li	Be											B	C	N	O	F	Ne
Na	Mg											Al	Si	P	S	Cl	Ar
K	Ca	Sc	Ti	V	Cr	Mn	Fe	Co	Ni	Cu	Zn	Ga	Ge	As	Se	Br	Kr
Rb	Sr	Y	Zr	Nb	Mo	Tc	Ru	Rh	Pd	Ag	Cd	In	Sn	Sb	Te	I	Xe
Cs	Ba	*La	Hf	Ta	W	Re	Os	Ir	Pt	Au	Hg	Tl	Pb	Bi	Po	At	Rn
Fr	Ra	†Ac	Rf	Db	Sg	Bh	Hs	Mt	Ds	Rg	Cn	Uut	Uuq	Uup	Uuh	Uus	Uuo

*	Ce	Pr	Nd	Pm	Sm	Eu	Gd	Tb	Dy	Ho	Er	Tm	Yb	Lu
†	Th	Pa	U	Np	Pu	Am	Cm	Bk	Cf	Es	Fm	Md	No	Lr

Dysprosium is a metal that ranks ninth in abundance among the rare earth elements in the Earth's crust. It was discovered in 1886 by the French chemist Paul-Émile Lecoq de Boisbaudran in a sample of erbium oxide. He based its name on the Greek word *dysprositos*, which means "hard to get at."

Although discovered in 1886, pure dysprosium was not actually available until 1950, when modern chemical techniques such as ion-exchange separation were developed.

The most important source of dysprosium is now China, which produces 97 percent of the world's rare earth supplies from mines in Inner Mongolia. Smaller quantities are found in monazite ores found in river sand in India, Brazil, and South Africa and bastnasite ores from the Mojave Desert in California, which, for economic reasons, are only being mined in small quantities. The dysprosium is isolated from its ores by ion exchange (see praseodymium).

Dysprosium metal resembles most of the other rare earth metals. It is soft enough to be cut with a knife; has a shiny, silvery color; and is relatively stable in air.

Some isotopes of dysprosium are effective absorbers of neutrons and are being considered for use in the control rods in nuclear reactors. Dysprosium is also used in color television tubes and mercury lamps. Like terbium, it forms alloys whose magnetic properties are used to help compact discs function more effectively.

IA	IA	IIIB	IVB	VB	VIB	VIIB	VIIIB			IB	IIB	IIIA	IVA	VA	VIA	VIIA	VIIIA
H																	He
Li	Be											B	C	N	O	F	Ne
Na	Mg											Al	Si	P	S	Cl	Ar
K	Ca	Sc	Ti	V	Cr	Mn	Fe	Co	Ni	Cu	Zn	Ga	Ge	As	Se	Br	Kr
Rb	Sr	Y	Zr	Nb	Mo	Tc	Ru	Rh	Pd	Ag	Cd	In	Sn	Sb	Te	I	Xe
Cs	Ba	*La	Hf	Ta	W	Re	Os	Ir	Pt	Au	Hg	Tl	Pb	Bi	Po	At	Rn
Fr	Ra	†Ac	Rf	Db	Sg	Bh	Hs	Mt	Ds	Rg	Cn	Uut	Uuq	Uup	Uuh	Uus	Uuo

*	Ce	Pr	Nd	Pm	Sm	Eu	Gd	Tb	Dy	Ho	Er	Tm	Yb	Lu
†	Th	Pa	U	Np	Pu	Am	Cm	Bk	Cf	Es	Fm	Md	No	Lr

Holmium

Holmium is a fairly scarce rare earth element. In 1878, two Swiss scientists noticed its characteristic spectral lines but could not identify them. They called the unknown source of the spectral lines element X. Soon afterward, in 1879, the Swedish chemist Per Teodor Cleve isolated and identified the element while working with a mineral called erbia. He named the new element for his native city of Stockholm, using the Latin version of its name, Holmia.

The most important commercial source of holmium is the Inner Mongolian mines in China. Monazite found in river sand in India, Brazil, and South Africa and bastnasite deposits found in the Mojave Desert in California also contain small quantities of holmium, but for economic reasons are not extensively mined. The holmium is isolated from its ores by ion exchange (see praseodymium).

Pure metallic holmium, which was not available until quite recently, resembles most rare earth metals. It has a bright silvery color and is malleable, ductile, and quite soft. It is fairly corrosion-resistant in dry air but tarnishes quickly in moist air, forming a yellowish oxide. The yellow oxide is often used as a coloring agent to color glass yellow.

Holmium laser treatment has become an important alternative to surgery for the treatment of an enlarged prostate. This condition, known as benign prostatic hyperplasia, is estimated to affect more than 80 percent of men over 80. Unlike surgery, the energy in a concentrated laser beam is used to remove tissue that is blocking the proper functioning of the urethra.

Atomic Number **67**

Chemical Symbol **Ho**

Group **IIIB—Rare Earth Element (Lanthanides)**

Erbium

IA	IA	IIIB	IVB	VB	VIB	VIIB	VIIIB			IB	IIB	IIIA	IVA	VA	VIA	VIIA	VIIIA
H																	He
Li	Be											B	C	N	O	F	Ne
Na	Mg											Al	Si	P	S	Cl	Ar
K	Ca	Sc	Ti	V	Cr	Mn	Fe	Co	Ni	Cu	Zn	Ga	Ge	As	Se	Br	Kr
Rb	Sr	Y	Zr	Nb	Mo	Tc	Ru	Rh	Pd	Ag	Cd	In	Sn	Sb	Te	I	Xe
Cs	Ba	*La	Hf	Ta	W	Re	Os	Ir	Pt	Au	Hg	Tl	Pb	Bi	Po	At	Rn
Fr	Ra	†Ac	Rf	Db	Sg	Bh	Hs	Mt	Ds	Rg	Cn	Uut	Uuq	Uup	Uuh	Uus	Uuo

*	Ce	Pr	Nd	Pm	Sm	Eu	Gd	Tb	Dy	Ho	Er	Tm	Yb	Lu
†	Th	Pa	U	Np	Pu	Am	Cm	Bk	Cf	Es	Fm	Md	No	Lr

Erbium is a fairly scarce rare earth metal. It was discovered in 1843 by Carl Gustaf Mosander in a yellow oxide that he isolated from the mineral called yttria. Mosander named the element for the Swedish village of Ytterby, the site of large concentrations of yttria and erbium.

China is today the dominant producer of the rare earths. Its mines in Inner Mongolia supply some 97 percent of the world's requirements. Smaller quantities are found in the minerals xenotine and euxerite. Erbium is actually an impurity in these ores. It is separated from the other rare earth elements by ion exchange (see praseodymium).

Erbium in its pure metallic form resembles most of the other rare earth metals. It is soft and malleable and has the usual silvery luster of these metals. By comparison with the other rare earths, it is fairly corrosion-resistant.

Erbium lasers are used in dermatology to remove wrinkles and blemishes on the surface of the skin. It is a mild treatment where the energy in the laser beam removes the outer layer of the skin to expose the smoother undamaged underlying layers of skin. And in the field of electronics erbium-doped fiber amplifiers are used to extend the range of light signals transmitted by fiber optics. They boost and reamplify the signal that would otherwise be attenuated over long distances.

The commercial applications of erbium are rather limited. It is used to form special alloys with metals such as vanadium in order to improve their workability and malleability. Erbium oxides are often added to glass and enamel glazes to color them pink. The glass is often used for sunglasses and inexpensive jewelry.

Atomic Number **68**

Chemical Symbol **Er**

Group **IIIB—Rare Earth Element (Lanthanides)**

Thulium

IA																	VIIIA
H	IA											IIIA	IVA	VA	VIA	VIIA	He
Li	Be											B	C	N	O	F	Ne
Na	Mg	IIIB	IVB	VB	VIB	VIIB		VIIIB		IB	IIB	Al	Si	P	S	Cl	Ar
K	Ca	Sc	Ti	V	Cr	Mn	Fe	Co	Ni	Cu	Zn	Ga	Ge	As	Se	Br	Kr
Rb	Sr	Y	Zr	Nb	Mo	Tc	Ru	Rh	Pd	Ag	Cd	In	Sn	Sb	Te	I	Xe
Cs	Ba	*La	Hf	Ta	W	Re	Os	Ir	Pt	Au	Hg	Tl	Pb	Bi	Po	At	Rn
Fr	Ra	†Ac	Rf	Db	Sg	Bh	Hs	Mt	Ds	Rg	Cn	Uut	Uuq	Uup	Uuh	Uus	Uuo

*	Ce	Pr	Nd	Pm	Sm	Eu	Gd	Tb	Dy	Ho	Er	Tm	Yb	Lu
†	Th	Pa	U	Np	Pu	Am	Cm	Bk	Cf	Es	Fm	Md	No	Lr

Tm

Thulium is a rare earth metal that is extremely scarce. It occurs in very small quantities in the company of other rare earths. The Swedish chemist Per Teodor Cleve discovered the element in 1879 by isolating its greenish oxide from the mineral erbia. He named his new element for Thule, the ancient name for Scandinavia.

The principal source of thulium is China's mines in Inner Mongolia, which supply some 97 percent of the world's rare earth elements. Smaller quantities are found in monazite and bastnasite ores. Monazite contains all of the rare elements and contains approximately 0.007 of 1 percent thulium. It is found in river sand in India, Brazil, and South Africa. Bastnasite deposits are found in the Mojave Desert in California, but for economic reasons are only being mined in small quantities. The neodymium is isolated from its ores by ion exchange (see praseodymium).

Like many of the rare earth metals, thulium is a bright silvery metal that is ductile, malleable, and soft enough to be cut with a knife. It is fairly resistant to corrosion in dry air. (It is expensive, and very little of the metal is available for experimentation.)

Thulium lasers have become quite widespread in medicine as an alternate procedure to surgery. They are used to remove polyps and to treat benign prostate enlargement in men. More recently they have found a place in dermatology to resurface damaged or wrinkled skin.

Atomic Number **69**

Chemical Symbol **Tm**

Group **IIIB—Rare Earth Element (Lanthanides)**

Ytterbium

IA																	VIIIA
H	IA											IIIA	IVA	VA	VIA	VIIA	He
Li	Be											B	C	N	O	F	Ne
Na	Mg	IIIB	IVB	VB	VIB	VIIB		VIIIB		IB	IIB	Al	Si	P	S	Cl	Ar
K	Ca	Sc	Ti	V	Cr	Mn	Fe	Co	Ni	Cu	Zn	Ga	Ge	As	Se	Br	Kr
Rb	Sr	Y	Zr	Nb	Mo	Tc	Ru	Rh	Pd	Ag	Cd	In	Sn	Sb	Te	I	Xe
Cs	Ba	*La	Hf	Ta	W	Re	Os	Ir	Pt	Au	Hg	Tl	Pb	Bi	Po	At	Rn
Fr	Ra	†Ac	Rf	Db	Sg	Bh	Hs	Mt	Ds	Rg	Cn	Uut	Uuq	Uup	Uuh	Uus	Uuo

*	Ce	Pr	Nd	Pm	Sm	Eu	Gd	Tb	Dy	Ho	Er	Tm	Yb	Lu
†	Th	Pa	U	Np	Pu	Am	Cm	Bk	Cf	Es	Fm	Md	No	Lr

Ytterbium is a rare earth metal found in modest abundance in the Earth's crust and always in the company of other rare earths. The French chemist Jean de Marignac discovered this element in 1878 as a component of the mineral known as erbia. Ytterbium, which Marignac named for the Swedish village of Ytterby on the basis of its high concentrations of erbium, was actually the first rare earth element to be discovered.

The principal source of ytterbium is the Inner Mongolian mines of China. Smaller quantities are found in monazite and bastnasite ores. Monazite, found in river sand in India, Brazil, and South Africa, contains ytterbium in concentrations of approximately 0.03 of 1 percent. Bastnasite deposits are found in the Mojave desert in California. The neodymium is isolated from its ores by ion exchange (see praseodymium).

Pure ytterbium metal was not available for study until 1953. Like so many of the rare earth metals, it has a bright silvery luster and is soft, malleable, and ductile. It oxidizes when exposed to air and is therefore usually stored in sealed containers.

Along with erbium, ytterbium has recently found use as an amplifier in fiber-optic cables, increasing the range that cables can transmit signals. Ytterbium-doped fibers are interspersed periodically along the length of the cable and, when stimulated by an optical signal passing through the cable, function as lasers boosting the strength of the signal.

Ytterbium fiber lasers are finding use in the marking of industrial products. Producing a laser beam in the infrared region of the electromagnetic spectrum, the ytterbium laser, using its heating ability, can efficiently etch identifying markings on products. The intense heat it can produce is also being used to weld and generally machine metals and composite materials.

Atomic Number **70**

Chemical Symbol **Yb**

Group IIIB—Rare Earth Element (Lanthanides)

IA																	VIIIA
H	IA											IIIA	IVA	VA	VIA	VIIA	He
Li	Be											B	C	N	O	F	Ne
Na	Mg	IIIB	IVB	VB	VIB	VIIB		VIIIB		IB	IIB	Al	Si	P	S	Cl	Ar
K	Ca	Sc	Ti	V	Cr	Mn	Fe	Co	Ni	Cu	Zn	Ga	Ge	As	Se	Br	Kr
Rb	Sr	Y	Zr	Nb	Mo	Tc	Ru	Rh	Pd	Ag	Cd	In	Sn	Sb	Te	I	Xe
Cs	Ba	*La	Hf	Ta	W	Re	Os	Ir	Pt	Au	Hg	Tl	Pb	Bi	Po	At	Rn
Fr	Ra	†Ac	Rf	Db	Sg	Bh	Hs	Mt	Ds	Rg	Cn	Uut	Uuq	Uup	Uuh	Uus	Uuo

*	Ce	Pr	Nd	Pm	Sm	Eu	Gd	Tb	Dy	Ho	Er	Tm	Yb	Lu
†	Th	Pa	U	Np	Pu	Am	Cm	Bk	Cf	Es	Fm	Md	No	Lr

Lutetium

Lu

Lutetium is one of the least abundant of the rare earth elements. It was discovered in 1907 by the Austrian mineralogist Baron Carl Auer von Welsbach and the French scientist Georges Urbain as an impurity in a mineral sample thought to contain only ytterbium. After some confusion, Urbain composed the name for his newly identified element from *Lutetia*, the ancient name for Paris. Welsbach had wanted *cassiopium*, a name taken from the constellation Cassiopeia. Many German scientists still refer to lutetium by the name *cassiopium*.

Although he never formally published his results, the American chemist Charles James, known to his colleagues and students as "King" James, is now considered by many scientists to have independently discovered lutetium in 1907. Working during the early 1900s at the University of New Hampshire, James became a major force in the production of rare earth elements. His work was recognized in 1999 by the American Chemical Society, and his laboratory, known today as Conant Hall, was designated a National Historic Chemical Landmark. Lutetium is thus the only naturally occurring element discovered in the United States.

The principal supply of lutetium comes from mines in Inner Mongolia, China, where some 97 percent of the world's supply of rare earths are produced. Monazite-bearing river sands, located chiefly in Brazil, India, and South Africa, also contribute small amounts. Monazite contains all the rare earth elements, and lutetium, present at a concentration of 0.003 of 1 percent, can be removed from it by ion exchange (see praseodymium). Pure lutetium metal is difficult and expensive to prepare. Like most of the other rare earth metals, it is silvery white and corrosion-resistant. As befits its position as the last lanthanide, it is the hardest and heaviest rare earth element.

Atomic Number **71**

Chemical Symbol **Lu**

Group **IIIB—Rare Earth Element (Lanthanides)**

Hafnium

IA	IA	IIIB	IVB	VB	VIB	VIIB	VIIIB	VIIIB	VIIIB	IB	IIB	IIIA	IVA	VA	VIA	VIIA	VIIIA
H																	He
Li	Be											B	C	N	O	F	Ne
Na	Mg											Al	Si	P	S	Cl	Ar
K	Ca	Sc	Ti	V	Cr	Mn	Fe	Co	Ni	Cu	Zn	Ga	Ge	As	Se	Br	Kr
Rb	Sr	Y	Zr	Nb	Mo	Tc	Ru	Rh	Pd	Ag	Cd	In	Sn	Sb	Te	I	Xe
Cs	Ba	*La	**Hf**	Ta	W	Re	Os	Ir	Pt	Au	Hg	Tl	Pb	Bi	Po	At	Rn
Fr	Ra	†Ac	Rf	Db	Sg	Bh	Hs	Mt	Ds	Rg	Cn	Uut	Uuq	Uup	Uuh	Uus	Uuo

*	Ce	Pr	Nd	Pm	Sm	Eu	Gd	Tb	Dy	Ho	Er	Tm	Yb	Lu
†	Th	Pa	U	Np	Pu	Am	Cm	Bk	Cf	Es	Fm	Md	No	Lr

Atomic Number 72

Chemical Symbol Hf

Group IVB—Transition Element

Hafnium is a bright, silvery-white metal that is extremely ductile. Its properties as well as its history are closely tied to the element zirconium. The two metals are so chemically similar that it is almost impossible to prepare one without traces of the other.

Although hafnium is a fairly abundant metal, it was not discovered until 1923. Numerous investigators, including Mendeleyev, had predicted the existence of element 72, but the omnipresence of its chemical twin zirconium interfered with its identification. Finally, Dirk Coster, a Dutch physicist, and George Karl von Hevesy, a Hungarian physicist, found the element using a prediction of its electron configuration based on the new quantum theory of electron shells and subshells developed by Niels Bohr. They used an X-ray analysis of the electronic structure of hafnium to discover it in zircon, the ore that is the principal source of zirconium metal. To honor Bohr, Coster and de Hevesy named their new element for his native city, Copenhagen, using the Latin form of its name, Hafnia.

Hafnium is extremely resistant to corrosion. It reacts with the oxygen in air to form a protective film that prevents its further oxidation. It is less dense than zirconium, but its chemistry is almost identical to that of zirconium. Hafnium is obtained commercially from zircon and baddeleyite ores. As with zirconium, its extraction is done with the Kroll process, in which these ores are first treated with chlorine gas to produce hafnium chloride, which is reduced to yield hafnium metal by the addition of either magnesium or sodium.

The principal use of hafnium is based on one of its few differences from zirconium. Unlike zirconium, hafnium is an excellent absorber of thermal neutrons. These are neutrons that have been slowed down by the so-called moderator in a nuclear reactor and are more easily absorbed by uranium-235 than are fast-moving neutrons. The ability of hafnium to absorb slow-moving neutrons has made it a most useful material for the construction of the reactor control rods that are lowered into the heart of a reactor to absorb neutrons. By removing neutrons, the rods can effectively slow or halt the chain reaction taking place in the reactor. The reactors used on nuclear submarines often use hafnium control rods. The main advantage of hafnium, compared to other common control rod materials such as cadmium and boron, is its strength and resistance to corrosion. Unfortunately, hafnium is rather expensive. In a fairly large reactor, one that may contain 40 to 50 control rods, the cost of the hafnium rods can be $1 million or more.

Other applications of hafnium are rather limited, but it is used to some extent in incandescent lamps and as a "getter" in many gas-filled systems. A hafnium "getter" removes—that is, "gets" rid of—unwanted gases such as oxygen and nitrogen from systems in which these gases are not wanted.

Many chemists had predicted the existence of element 72, but the omnipresence of its chemical twin zirconium interfered with its identification. It was not discovered until 1923.

Tungsten

IA	IA	IIIB	IVB	VB	VIB	VIIB	VIIIB			IB	IIB	IIIA	IVA	VA	VIA	VIIA	VIIIA
H																	He
Li	Be											B	C	N	O	F	Ne
Na	Mg											Al	Si	P	S	Cl	Ar
K	Ca	Sc	Ti	V	Cr	Mn	Fe	Co	Ni	Cu	Zn	Ga	Ge	As	Se	Br	Kr
Rb	Sr	Y	Zr	Nb	Mo	Tc	Ru	Rh	Pd	Ag	Cd	In	Sn	Sb	Te	I	Xe
Cs	Ba	*La	Hf	Ta	W	Re	Os	Ir	Pt	Au	Hg	Tl	Pb	Bi	Po	At	Rn
Fr	Ra	†Ac	Rf	Db	Sg	Bh	Hs	Mt	Ds	Rg	Cn	Uut	Uuq	Uup	Uuh	Uus	Uuo

*	Ce	Pr	Nd	Pm	Sm	Eu	Gd	Tb	Dy	Ho	Er	Tm	Yb	Lu
†	Th	Pa	U	Np	Pu	Am	Cm	Bk	Cf	Es	Fm	Md	No	Lr

Atomic Number **74**

Chemical Symbol **W**

Group VIB—Transition Element

W

In its raw form, tungsten is a steel-gray metal that is often fairly brittle and hard to work. Yet if all of its impurities are removed, it is soft enough to be cut with a sharp saw. The history of chemistry during the 18th century is often difficult to disentangle, but it seems that two Spanish brothers, Juan José and Fausto de Elhuar, discovered tungsten in 1783, isolating it from the mineral known as wolframite. There is evidence, however, that the element was known before that time and was called wolfram. Many chemists, particularly in Germany, still refer to tungsten as wolfram. Moreover, the chemical symbol for tungsten is W and is taken from that name. The more widely accepted name for the element is taken from the Swedish words *tung stem*, which mean "heavy stone."

The principal ores of tungsten are wolframite and scheelite. Approximately 75 percent of the world's tungsten resources are thought to exist in China.

One of the most important uses of tungsten is in the manufacture of filaments for the common light bulb. Tungsten has the highest melting point (3,410°C) and highest boiling point (5,900°C) of any metal. The high-temperature applications of tungsten range from heating elements in electric heaters to the nozzles on the rocket motors of space vehicles. Electricity flowing through a coiled wire of tungsten produces enough heat to make the wire white hot. To prevent the metal from overheating, inert gases such as nitrogen and argon are enclosed in the bulb containing a tungsten filament.

Even though tungsten has the lowest vapor pressure of any metal, some of it does slowly vaporize with use, forming a dark

Because tungsten has the highest melting point of any metal, it is used in the filaments of ordinary light bulbs.

deposit on the insides of bulbs. Tungsten filaments are also used in television tubes and in the cathode ray tubes used in computer monitors.

Tungsten is often added to steel to form an alloy called tungsten steel. Cutting tools made of common carbon steel are often unable to hold their edge at high temperatures. By contrast, tools made of tungsten steel remain sharp even under red-hot temperature conditions.

The compound known as tungsten carbide is extremely hard and chemically inert even at very high temperatures. It finds important industrial use in the manufacture of high-speed cutting tools.

Rhenium

IA	IA	IIIB	IVB	VB	VIB	VIIB	VIIIB			IB	IIB	IIIA	IVA	VA	VIA	VIIA	VIIIA
H																	He
Li	Be											B	C	N	O	F	Ne
Na	Mg											Al	Si	P	S	Cl	Ar
K	Ca	Sc	Ti	V	Cr	Mn	Fe	Co	Ni	Cu	Zn	Ga	Ge	As	Se	Br	Kr
Rb	Sr	Y	Zr	Nb	Mo	Tc	Ru	Rh	Pd	Ag	Cd	In	Sn	Sb	Te	I	Xe
Cs	Ba	*La	Hf	Ta	W	Re	Os	Ir	Pt	Au	Hg	Tl	Pb	Bi	Po	At	Rn
Fr	Ra	†Ac	Rf	Db	Sg	Bh	Hs	Mt	Ds	Rg	Cn	Uut	Uuq	Uup	Uuh	Uus	Uuo

*	Ce	Pr	Nd	Pm	Sm	Eu	Gd	Tb	Dy	Ho	Er	Tm	Yb	Lu
†	Th	Pa	U	Np	Pu	Am	Cm	Bk	Cf	Es	Fm	Md	No	Lr

Re

Rhenium is one of the rarest of elements. Early investigators during the latter part of the 1920s worked through approximately 1 million pounds of the mineral molybdenite to recover 1 gram of rhenium. Because of the difficulty in extracting it, the price of the element rose to an extraordinary $10,000 per gram in 1928. With more efficient extraction techniques, rhenium has become considerably less expensive today.

Rhenium is an extremely dense metal with a silvery-gray luster and a melting point exceeded only by those of tungsten and carbon. It was discovered in platinum ores in 1925 by the German chemists Ida Tacke, Walter Nodack, and Otto Carl Berg. Illustrating the power of the periodic table, they knew that element 75 should fill a gap in column VIIB of the table and were able to predict many of the reactions of this element even before its discovery. They named the element for the Rhine River, using the Latin version of its name, Rhenus.

The principal commercial sources of rhenium are the ores molybdenite and copper sulfide. Approximately 16,000 pounds of rhenium are produced each year at the Sierrita mine near Tucson, Arizona, the only source of rhenium in the United States.

Rhenium is chiefly used as an alloying agent for fabricating metals that are resistant to wear and have good high-temperature capability. Examples are electrical switch contacts, electrodes, turbine blades for jet aircraft, and iridium-coated rhenium that lines the interior of the combustion chamber in jet engines. The high melting point of rhenium (3,180°C) is also the basis for its use in combination with tungsten to make thermocouples for measuring temperatures as high as 2,000°C. Another important use of rhenium when alloyed with platinum is as a catalyst for the production of lead-free gasoline.

Atomic Number 75

Chemical Symbol Re

Group VIIB—Transition Element

IA																	VIIIA
H	IA											IIIA	IVA	VA	VIA	VIIA	He
Li	Be											B	C	N	O	F	Ne
Na	Mg	IIIB	IVB	VB	VIB	VIIB	VIIIB			IB	IIB	Al	Si	P	S	Cl	Ar
K	Ca	Sc	Ti	V	Cr	Mn	Fe	Co	Ni	Cu	Zn	Ga	Ge	As	Se	Br	Kr
Rb	Sr	Y	Zr	Nb	Mo	Tc	Ru	Rh	Pd	Ag	Cd	In	Sn	Sb	Te	I	Xe
Cs	Ba	*La	Hf	Ta	W	Re	**Os**	Ir	Pt	Au	Hg	Tl	Pb	Bi	Po	At	Rn
Fr	Ra	†Ac	Rf	Db	Sg	Bh	Hs	Mt	Ds	Rg	Cn	Uut	Uuq	Uup	Uuh	Uus	Uuo

*	Ce	Pr	Nd	Pm	Sm	Eu	Gd	Tb	Dy	Ho	Er	Tm	Yb	Lu
†	Th	Pa	U	Np	Pu	Am	Cm	Bk	Cf	Es	Fm	Md	No	Lr

Osmium

Atomic Number **76**

Chemical Symbol **Os**

Group **VIIIB—Transition Element**

Os

Osmium is a hard, brittle, bluish-white metal with an extremely high melting point (3,054°C). Because the pure metal is difficult to make, it is often fabricated as a powder, which is then formed into a solid mass by heating. The powder oxidizes in air to form osmium tetroxide (OsO_4), which is slowly emitted as a strong-smelling toxic gas capable of causing lung and skin damage.

Osmium was discovered in 1803 by the English chemist Smithson Tennant, who isolated it by examining the residue formed by treating platinum ores with aqua regia, a mixture of nitric and hydrochloric acids. Tennant named the newly discovered element for its noxious odor, using the Greek word *osme*, which means "smell."

Osmium is chiefly found in nickel- and platinum-bearing ores. The difficulty in refining it from these ores is offset by the platinum and nickel that are also recovered.

The emission of its poisonous oxide gas makes the use of osmium metal impractical. As an alloying additive, however, it is quite safe and is chiefly used to make hard alloys with such metals as platinum and iridium. These alloys are used for electrical switch contacts, phonograph needles, and fountain-pen tips. Furthermore, although it is dangerous, osmium tetroxide in very dilute solutions is widely used to stain substances to be viewed on microscope slides.

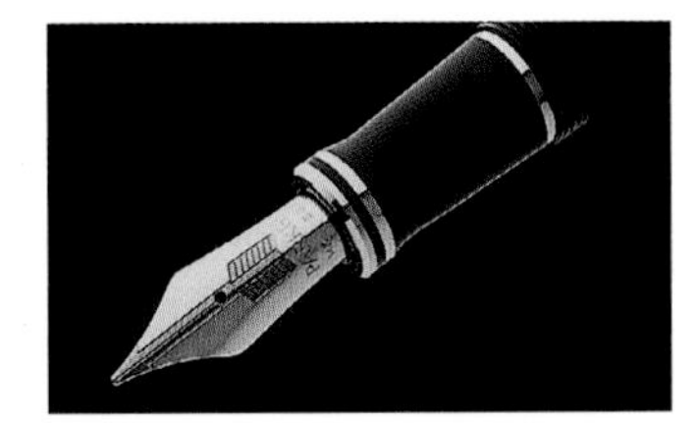

Osmium's hardness makes it an ideal ingredient for the alloy used to make fountain-pen tips.

Iridium

IA	IA	IIIB	IVB	VB	VIB	VIIB	VIIIB			IB	IIB	IIIA	IVA	VA	VIA	VIIA	VIIIA
H																	He
Li	Be											B	C	N	O	F	Ne
Na	Mg											Al	Si	P	S	Cl	Ar
K	Ca	Sc	Ti	V	Cr	Mn	Fe	Co	Ni	Cu	Zn	Ga	Ge	As	Se	Br	Kr
Rb	Sr	Y	Zr	Nb	Mo	Tc	Ru	Rh	Pd	Ag	Cd	In	Sn	Sb	Te	I	Xe
Cs	Ba	*La	Hf	Ta	W	Re	Os	Ir	Pt	Au	Hg	Tl	Pb	Bi	Po	At	Rn
Fr	Ra	†Ac	Rf	Db	Sg	Bh	Hs	Mt	Ds	Rg	Cn	Uut	Uuq	Uup	Uuh	Uus	Uuo

*	Ce	Pr	Nd	Pm	Sm	Eu	Gd	Tb	Dy	Ho	Er	Tm	Yb	Lu
†	Th	Pa	U	Np	Pu	Am	Cm	Bk	Cf	Es	Fm	Md	No	Lr

Ir

Iridium is one of the hardest and most corrosion-resistant metals known. It is considered a precious metal, similar to platinum. It is a yellowish-white metal that is more than 20 times denser than water, and it is too brittle to machine easily or form into specific shapes by compression or extension. The English chemist Smithson Tennant discovered iridium in 1803, using the same procedure that led him to the discovery of osmium. He isolated the element in the residue of platinum ores treated with aqua regia, a mixture of nitric and hydrochloric acids. Tennant named it iridium from the Latin word *iris*, meaning "rainbow," because its salts are highly colored.

Iridium is generally found in ores containing platinum or nickel. Separating it from these ores is a laborious and costly task that is justified only by the simultaneous recovery of valuable platinum and nickel.

The chief application of iridium is as an additive to platinum, creating alloys that increase the hardness of the latter metal. Its resistance to corrosion has made it useful in the fabrication of items that require absolute purity such as hypodermic needles and rocket engines. The stability of iridium led to its being used to fabricate the platinum bar whose length was defined as the standard meter. Kept in Paris, the bar consisted of an alloy of 90 percent platinum and 10 percent iridium. The bar has since been replaced by a standard based on the wavelength of the natural vibrations of the krypton atom.

Atomic Number 77

Chemical Element Ir

Group VIIIB—Transition Element

IA	IA	IIIB	IVB	VB	VIB	VIIB	VIIIB			IB	IIB	IIIA	IVA	VA	VIA	VIIA	VIIIA
H																	He
Li	Be											B	C	N	O	F	Ne
Na	Mg											Al	Si	P	S	Cl	Ar
K	Ca	Sc	Ti	V	Cr	Mn	Fe	Co	Ni	Cu	Zn	Ga	Ge	As	Se	Br	Kr
Rb	Sr	Y	Zr	Nb	Mo	Tc	Ru	Rh	Pd	Ag	Cd	In	Sn	Sb	Te	I	Xe
Cs	Ba	*La	Hf	Ta	W	Re	Os	Ir	Pt	Au	Hg	Tl	Pb	Bi	Po	At	Rn
Fr	Ra	†Ac	Rf	Db	Sg	Bh	Hs	Mt	Ds	Rg	Cn	Uut	Uuq	Uup	Uuh	Uus	Uuo

*	Ce	Pr	Nd	Pm	Sm	Eu	Gd	Tb	Dy	Ho	Er	Tm	Yb	Lu
†	Th	Pa	U	Np	Pu	Am	Cm	Bk	Cf	Es	Fm	Md	No	Lr

Atomic Number **78**

Chemical Symbol **Pt**

Group **VIIIB—Transition Element** (Precious Metal)

Pt

Platinum is a precious metal that, like gold, is highly prized in the making of jewelry. Platinum is also a major industrial metal that is actively traded in commodity markets throughout the world. Its current price can be determined from any financial newspaper. It has been estimated that one of every five products made today uses platinum, either directly in the product itself or as part of the manufacturing process for the product.

Platinum is a durable, heavy, silvery-white metal that is malleable, or easily rolled or hammered into thin sheets, and is

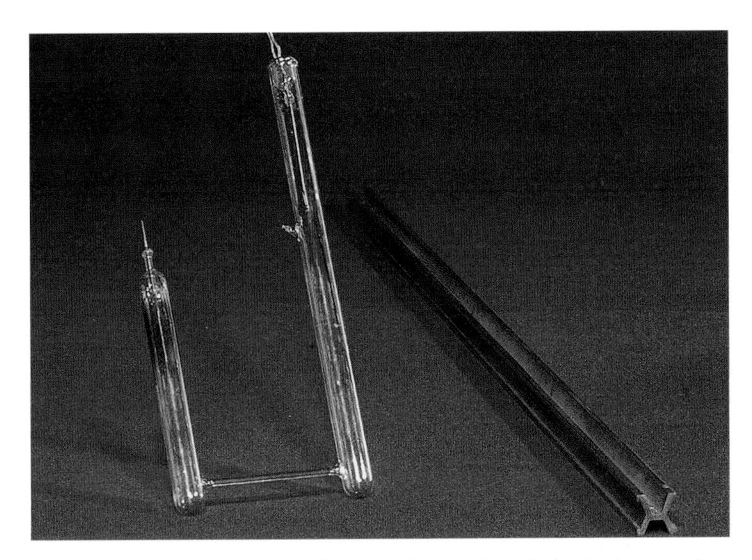

From 1889 until 1960, two marks on the bar on the right, a platinum and iridium alloy, were the standard used to define the meter. The bar was kept in France by the Bureau of International Weights and Measures. The meter is now defined using a krypton lamp (left).

Platinum is a major industrial metal that is actively traded in commodity markets throughout the world.

extremely resistant to corrosion. It does not react with the oxygen in air and is therefore usually found in nature as a pure metal. Retrieving the metal is more difficult than simply picking nuggets of it off the ground, however, because other metals are usually mixed with platinum in its ore. Usually, some form of industrial and chemical "cleanup" is needed to recover pure platinum.

More than 80 tons of platinum are produced each year. Although fairly large deposits of the metal exist in the Ural Mountains in Russia, most of the world's supply comes from the Republic of South Africa.

Platinum was discovered by Charles Wood, an Englishman, in 1741. There is evidence, however, that it was used by the pre-Columbian Indians of the Americas. It takes its name from the Spanish word *platina*, which means "silver."

Platinum is used in petroleum refining, dentistry, the chemical industry, the ceramics industry, and the electrical and electronics industries. In recent years, the automobile industry has found a major use for the metal. A platinum-coated ceramic grid serves as the catalyst in the catalytic converter attached to the exhaust system of all automobiles manufactured in the United States. The platinum assists chemical reactions that "clean up" the exhaust gases coming from the engine of the car, converting carbon monoxide and unburned fuel into water and carbon dioxide.

Many uses of platinum take advantage of its chemical stability and inertness. A bar of a platinum-iridium alloy, for example, is stored in a special vault in Paris, where it serves as the world's standard for the kilogram, the basic unit of mass in the metric system. Platinum alloys are also used in the glass industry to contain and handle molten glass, and because the rate at which platinum expands upon heating is almost the same as that of glass, platinum wires can be sealed into glass and used to conduct electricity through glass tubes and bulbs. In medicine, stable platinum electrodes are used in the electronic pacemakers that are implanted in the bodies of people with heart ailments to ensure that the heart maintains a steady rhythm.

Its ability to be hammered into incredibly thin sheets, some only 100 atoms thick, makes platinum ideal for thin protective coatings. It is used to coat the nose cones of missiles, the fuel nozzles of jet engines, and even the cutting edges of razor blades.

The catalytic converter used in automobiles is only one example of how effective platinum is as a catalyst for many different types of chemical reactions. As an example of the catalytic power of platinum, a mixture of hydrogen and oxygen will explode in the presence of the metal, and platinum is used to sustain the reaction in the hydrogen–oxygen fuel cells used in spacecraft. Platinum is often ground up into a fine powder to increase both its surface area and its ability to interact with other chemicals. It is commonly used in this form on airplanes that fly at high altitudes, which employ platinum catalysts in their air filters for converting the harmful ozone found in the upper layers of the atmosphere into oxygen. Many industrial processes depend on platinum catalysts. An example is the famous Ostwald process for making nitric acid, in which ammonia is burned in the presence of a platinum catalyst to produce nitric oxide. Platinum catalysts are also used to increase the rate of hydrogenation of fats and oils and the distillation of petroleum products from crude oil.

The International Bureau of Weights and Measures uses this piece of a platinum and iridium alloy as the official standard for the kilogram. The bar is kept in a vault at the organization's headquarters in France.

Recent research has shown that certain platinum compounds inhibit the growth of several kinds of cancerous tumors. The medical reason for this is not known, but progress has been reported in the use of platinum compounds to treat ovarian and testicular tumors, which are difficult to treat by more traditional methods.

Gold

IA	IA	IIIB	IVB	VB	VIB	VIIB	VIIIB			IB	IIB	IIIA	IVA	VA	VIA	VIIA	VIIIA
H																	He
Li	Be											B	C	N	O	F	Ne
Na	Mg											Al	Si	P	S	Cl	Ar
K	Ca	Sc	Ti	V	Cr	Mn	Fe	Co	Ni	Cu	Zn	Ga	Ge	As	Se	Br	Kr
Rb	Sr	Y	Zr	Nb	Mo	Tc	Ru	Rh	Pd	Ag	Cd	In	Sn	Sb	Te	I	Xe
Cs	Ba	*La	Hf	Ta	W	Re	Os	Ir	Pt	Au	Hg	Tl	Pb	Bi	Po	At	Rn
Fr	Ra	†Ac	Rf	Db	Sg	Bh	Hs	Mt	Ds	Rg	Cn	Uut	Uuq	Uup	Uuh	Uus	Uuo

*	Ce	Pr	Nd	Pm	Sm	Eu	Gd	Tb	Dy	Ho	Er	Tm	Yb	Lu
†	Th	Pa	U	Np	Pu	Am	Cm	Bk	Cf	Es	Fm	Md	No	Lr

Gold is considered the most precious of metals. Its value as money, in the form of gold coins and bars, is well known. It is traded in commodities exchanges, and the fluctuations in its price are considered an index of the health of the economy. Many countries use it as a standard for their monetary systems.

In its pure state, gold is one of the most beautiful of elements. With its bright yellow color and resistance to corrosion, it has been highly prized since ancient times. Its chemical symbol comes from the Latin word *aurum*, which means "shining dawn."

Gold is one of the most valuable and precious metals. The discovery of gold in California in 1849 attracted thousands of people who hoped to strike it rich.

Atomic Number **79**

Chemical Symbol **Au**

Group **IB—Transition Element** (Precious Metal)

Its name has even older roots, coming from the Sanskrit word *jval* and the Anglo-Saxon word *gold*, both meaning "gold."

In nature, gold is usually found as a pure metal, most often in the form of nuggets or flakes. It also occurs in veins of pure gold or deposits of a class of minerals called tellurides. It is fairly widely distributed throughout the world and is always found near deposits of quartz and pyrite. Most of the world's supply of gold comes from South Africa. About two-thirds of the gold produced in the United States comes from Nevada and South Dakota. Gold is also found in seawater but in quantities too small to recover profitably.

The legendary gold rushes of California and Alaska have become an important part of the early history of the American West. The technique of panning for gold depends on its density, which is approximately nine times greater than that of sand or gravel. When the forty-niners swirled gold-bearing sand around in their shallow pans, the sand and gravel were washed over the rim, leaving the heavier gold dust behind.

Gold can also be recovered from its ore by treating the ore with mercury. Gold dissolves in mercury to form an amalgam, or alloy of mercury with another metal, that is similar to the amalgam used by dentists to fill teeth. The gold is then recovered by heating the amalgam and distilling away the mercury. Still another method of recovering gold is to treat its ore with a cyanide solution. Large ponds of cyanide solution are used in the extraction process. The gold dissolves in this solution to form a gold–cyanide ion. This solution is then separated from the residual ore by filtration, and the gold is recovered by treating the gold–cyanide with a more active metal, such as zinc, which displaces it from its complex with cyanide.

Gold can sustain its wonderful luster because it is one of the most unreactive of metals. Even concentrated nitric acid will not attack it. Gold will, however, dissolve very slowly in the solution of hydrochloric and nitric acids that was called aqua regia by the alchemists because of its ability to dissolve the king of metals.

Solid bars of gold. Because gold is one of the most unreactive of metals, it can sustain its brilliant luster.

Gold is very soft and is the most ductile and malleable of all metals. The gold leaf used for decoration is made by pounding

Modern hotels with large windows use glass coated with a thin film of gold to counter the greenhouse effect and prevent their lobbies from becoming overheated.

the metal into very thin sheets, which are often so thin that light can be seen through them. It has been estimated that 1 ounce of gold can be flattened to cover an area of 300 square feet. A stack of 10,000 gold leaves is only 1 millimeter thick.

Gold's resistance to corrosion and its ability to reflect infrared radiation and prevent excessive heating make it an excellent coating for space vehicles. Modern hotels with large windows use glass coated with a thin film of gold to counter the greenhouse effect and prevent their lobbies from becoming overheated. Dentistry and electronics are two other areas in which the chemical inertness of gold is of great value. Gold teeth can last for decades, and gold-coated switches and connectors remain efficient after years of electrical sparking, which always accompanies the opening and closing of switches.

The purity of gold is measured in carats. Pure gold is said to be 24-carat gold. Because gold is very soft, however, most gold jewelry, such as rings and necklaces, is made of 18-carat gold. This means that the object is only 75 percent gold. The remainder of the alloy is usually a metal such as nickel or copper added to the gold to harden it. The "white gold" used in jewelry is an alloy of nickel and gold. A radioactive isotope of gold, with a relatively short half-life of 2.7 days, is used for treating cancer. Known as gold-198, it is easily made by subjecting natural gold to a beam of neutrons in a nuclear reactor.

Gold has proven to be an important element in the developing field of nanotechnology, where nanoparticles of the element, too small to be seen with the naked eye, are created and used in such fields as electronics and medicine. The actual size of these particles is in the range of nanometers, that is, billionths of a meter, and thousands of times smaller than the diameter of a human hair.

Nanoparticles have optical and chemical properties that can be quite different from the usual properties of the element. Nanoparticles of gold, for example, are a deep red color and, depending on their shape, which can be dots, rods, or even stars, can be highly reactive. They are now being used in industry as a catalyst and in medicine as a treatment for cancer. Gold nanoparticles introduced in the body have been shown to preferentially seek out cancer cells. When coated with the proper medication, or when treated externally with relatively harmless infrared radiation to heat the gold particles, they can destroy the malignant cells while leaving healthy tissue unharmed.

Mercury

IA	IA	IIIB	IVB	VB	VIB	VIIB	VIIIB			IB	IIB	IIIA	IVA	VA	VIA	VIIA	VIIIA
H																	He
Li	Be											B	C	N	O	F	Ne
Na	Mg											Al	Si	P	S	Cl	Ar
K	Ca	Sc	Ti	V	Cr	Mn	Fe	Co	Ni	Cu	Zn	Ga	Ge	As	Se	Br	Kr
Rb	Sr	Y	Zr	Nb	Mo	Tc	Ru	Rh	Pd	Ag	Cd	In	Sn	Sb	Te	I	Xe
Cs	Ba	*La	Hf	Ta	W	Re	Os	Ir	Pt	Au	Hg	Tl	Pb	Bi	Po	At	Rn
Fr	Ra	†Ac	Rf	Db	Sg	Bh	Hs	Mt	Ds	Rg	Cn	Uut	Uuq	Uup	Uuh	Uus	Uuo

*	Ce	Pr	Nd	Pm	Sm	Eu	Gd	Tb	Dy	Ho	Er	Tm	Yb	Lu
†	Th	Pa	U	Np	Pu	Am	Cm	Bk	Cf	Es	Fm	Md	No	Lr

Atomic Number **80**

Chemical Symbol **Hg**

Group **IIB—Transition Element**

Mercury is the only metal that is liquid at room temperature. It is an extremely heavy metal with a silvery-white color that accounts for the name quicksilver that is often given to it. It is a fairly good conductor of electricity but, unlike most metals, is a rather poor conductor of heat.

Mercury is rarely found as a pure metal in nature. Its chief ore is cinnabar, a bright red mineral that is also called vermillion. Cinnabar is composed of mercury sulfide and is found chiefly in Spain and Italy. Spain has mercury mines that have been operating continuously for the past 2,000 years. Metallic mercury is recovered from cinnabar by heating the ore in air and condensing the mercury vapor.

Mercury was well known to the ancient Chinese, and samples of the element have been found in Egyptian tombs dating back to 1500 B.C. It takes its name from the planet Mercury and its chemical symbol from the Latin word *hydragyrum*, which means "liquid silver."

Mercury freezes at −38.9°C and boils at 357°C, making it a liquid over a very wide and convenient range of temperatures. This accounts for its many uses in homes and scientific laboratories. Some common household products that contain mercury are thermometers, barometers, themostats, silent wall switches, and fluorescent bulbs. Industrial applications of mercury include diffusion pumps that are used to produce powerful vacuums by efficiently evacuating the air from systems and mercury vapor lamps that generate the bluish-white light from streetlights.

An interesting property of mercury is its high surface tension. If, for example, you break a thermometer, the mercury that escapes from it forms little balls that roll around without adhering to anything and are hard to collect. A more useful property of mercury is its ability to dissolve other metals to form alloys known as amalgams. Dentists often use a silver–mercury amalgam to fill teeth. A silver filling actually consists of powdered silver dissolved in mercury. Gold also dissolves in mercury, and mercury is used to recover gold from its ore. The resulting gold–mercury amalgam is then heated until the mercury vaporizes, leaving behind the pure metallic gold. The mercury vapor can be condensed and reused.

Mercuric nitrate was once used in the manufacture of felt hats. Workers in contact with this compound gradually began to develop serious problems—including loss of memory. This led to the common expression "mad as a hatter" for bizarre behavior.

Mercury is an extremely toxic element. Like many of the heavy metals, it combines chemically with enzymes in the body, causing them to lose their ability to act as catalysts for vital body functions. It is easily absorbed into the body from the gastrointestinal tract and can even enter the body through the skin. Because mercury is quite volatile, care must also be taken to avoid breathing its vapors even at room temperature. The vapor is much more harmful than the metal itself. It has been estimated that the vapors from a quantity of mercury as small as a teaspoon would saturate a fairly large room within a week and make it unsafe to work in.

Mercury is a major hazard for chemists working in laboratories, since small amounts are inevitably spilled during many chemical procedures. The mercury usually ends up in small cracks in the floor or laboratory bench and is difficult to remove. The usual technique to lessen the danger is to sprinkle sulfur on the mercury. Not only does this reduce the production of mercury vapor, but also the sulfur combines with the mercury to form a sulfide which is less dangerous.

Mercury acts as a cumulative poison, which means that small amounts absorbed over a long period build up in the body and can eventually become hazardous. A compound of mercury, mercuric nitrate, was once used in the manufacture of felt hats. Workers in contact with this compound gradually began to develop serious problems, including the loss of hair and teeth, loss of memory, and a general deterioration of the nervous system. This led to the common expression "mad as a hatter" for bizarre behavior.

The very properties that make mercury poisonous to humans make it effective in dealing with insect pests. Mercuric chloride ($HgCl_2$), for example, known commercially as corrosive sublimate, is a poison used as a fungicide and pesticide. Mercurous chloride ($HgCl$) is not quite as soluble as the mercuric form of the chloride and so is not quite as toxic. It is called calomel and is used in agriculture to control root maggots and other pests on tubers and bulbs. Although no longer used in medicine, calomel was once used as a purgative and as a treatment for syphilis. Because compounds of mercury are so damaging, a major effort is currently under way to eliminate all sources of pollution by this element. Many mercury-containing compounds have been banned in industry and agriculture.

The extent of the problem involving the contamination of the environment with mercury became clear when it was discovered that certain microorganisms living in lakes and rivers can actually metabolize mercury. Mercury-containing compounds that were discharged into these waters were taken up by these micro-organisms and chemically changed to methylmercury compounds. Fish feeding on these microorganisms then began to accumulate these compounds in their tissues. By the time a fairly large fish came to the market for human consumption, the concentration of mercury that had built up in the fish was estimated to be as large as 40,000 to 50,000 times the concentration of mercury in the water itself. Major efforts are now underway to curtail the discharge of mercury-containing compounds into rivers and lakes and to try to clean up areas where damage has already been done.

Although not generally considered a poison, an unusual compound of mercury known as mercury fulminate is extremely dangerous in another way because it is an unstable explosive. Commercially, it is used for making blasting caps.

Mercury batteries have become quite common in many portable electronic devices. The mercury battery consists of a zinc anode and a mercuric oxide cathode. The battery develops a voltage of about 1.35 volts and has the advantage of providing a constant voltage even as it ages.

Mercury is sold and traded on world markets in units called "flasks." One flask is equal to 76 pounds.

Thallium

IA	IA	IIIB	IVB	VB	VIB	VIIB	VIIIB			IB	IIB	IIIA	IVA	VA	VIA	VIIA	VIIIA
H																	He
Li	Be											B	C	N	O	F	Ne
Na	Mg											Al	Si	P	S	Cl	Ar
K	Ca	Sc	Ti	V	Cr	Mn	Fe	Co	Ni	Cu	Zn	Ga	Ge	As	Se	Br	Kr
Rb	Sr	Y	Zr	Nb	Mo	Tc	Ru	Rh	Pd	Ag	Cd	In	Sn	Sb	Te	I	Xe
Cs	Ba	*La	Hf	Ta	W	Re	Os	Ir	Pt	Au	Hg	**Tl**	Pb	Bi	Po	At	Rn
Fr	Ra	†Ac	Rf	Db	Sg	Bh	Hs	Mt	Ds	Rg	Cn	Uut	Uuq	Uup	Uuh	Uus	Uuo

*	Ce	Pr	Nd	Pm	Sm	Eu	Gd	Tb	Dy	Ho	Er	Tm	Yb	Lu
†	Th	Pa	U	Np	Pu	Am	Cm	Bk	Cf	Es	Fm	Md	No	Lr

Tl

Thallium is a soft, heavy metal that resembles lead in appearance. Unlike lead, it is soft enough to be cut with a knife. It is also very malleable. It was discovered in 1861 by Sir William Crookes, who identified it by the brilliant green spectral line of its emitted light. The element was named for this line, using the Greek word *thallos*, which means "a green twig" or "shoot."

Thallium is a rather scarce element. It is found sparsely distributed in several mineral ores such as crookside, lorandite, and hutchinsonite and also in manganese nodules distributed on the floor of the ocean. Since the extraction of thallium from these ores can be difficult, a more common source of the metal is as a by-product of lead and zinc refining.

Thallium is quite active and slowly corrodes when exposed to air. It reacts with the oxygen in the air to form a heavy gray oxide, which eventually flakes off to expose a fresh surface of the metal that is subject to further oxidation.

Thallium and its compounds are extremely toxic, and there is also evidence that thallium is carcinogenic. Even its contact with the skin can be dangerous, although in extremely low concentrations thallium has been used to treat skin disorders such as ringworm. Thallium sulfate is an odorless and tasteless poison that was formerly used to kill rats and insects, but it has now been banned in the United States.

Neither thallium nor its compounds have many commercial applications. The electrical conductivity of some compounds of thallium, such as thallium sulfide, changes when they are exposed to infrared radiation, and this property has made these compounds useful in some types of photoelectric cells and infrared detectors.

Atomic Number 81

Chemical Symbol Tl

Group IIIA—Post-transition Metal

A radioactive isotope of thallium, thallium-201, is now being used to diagnose various types of disease. The isotope has a half-life of only 72.9 hours, so that it is quickly eliminated from the body. Also of great importance is the fact that when thallium decays it emits very penetrating gamma rays that can be detected outside the body.

The technique depends on the "binding" of thallium-201 to the muscle tissue of the heart. This binding will occur, however, only if the tissue receives an adequate supply of blood. If the blood supply is restricted because of a narrowing or blockage of an artery, for example, the tissues supplied by the artery will not take up the thallium-201.

The radioisotope is usually administered first while the patient is at rest and then after the patient is subjected to a period of exercise. The patient is scanned with a scintillation detector, a kind of "gamma ray camera" that can detect and identify gamma rays, both before and after the exertion. The resulting data are then fed into a computer that produces an image on screen. The physician can then compare the thallium-201 uptake before and after the exercise. An area of the heart that has impaired blood flow, for example, will show up on the screen as a dark spot.

Thallium was discovered in 1861 by Sir William Crookes, who identified it by the brilliant green spectral line of its emitted light.

Lead

IA	IA	IIIB	IVB	VB	VIB	VIIB	VIIIB			IB	IIB	IIIA	IVA	VA	VIA	VIIA	VIIIA
H																	He
Li	Be											B	C	N	O	F	Ne
Na	Mg											Al	Si	P	S	Cl	Ar
K	Ca	Sc	Ti	V	Cr	Mn	Fe	Co	Ni	Cu	Zn	Ga	Ge	As	Se	Br	Kr
Rb	Sr	Y	Zr	Nb	Mo	Tc	Ru	Rh	Pd	Ag	Cd	In	Sn	Sb	Te	I	Xe
Cs	Ba	*La	Hf	Ta	W	Re	Os	Ir	Pt	Au	Hg	Tl	Pb	Bi	Po	At	Rn
Fr	Ra	†Ac	Rf	Db	Sg	Bh	Hs	Mt	Ds	Rg	Cn	Uut	Uuq	Uup	Uuh	Uus	Uuo

*	Ce	Pr	Nd	Pm	Sm	Eu	Gd	Tb	Dy	Ho	Er	Tm	Yb	Lu
†	Th	Pa	U	Np	Pu	Am	Cm	Bk	Cf	Es	Fm	Md	No	Lr

Lead is a relatively soft, dull-gray metal that is highly malleable and can be easily worked to make utensils of all kinds. It is a poor conductor of electricity. Lead has an ancient history. Lead coins and sculpture have been found in Egyptian tombs dating back to 5000 B.C., and lead pipes and plumbing used by the Romans can still be found in Italy today. The alchemists associated lead with the planet Saturn and not only believed lead to be the oldest metal but also believed that all other metals eventually transformed themselves into lead. It is also mentioned in the Bible in Exodus and the Book of Job. The element takes its name from the Anglo-Saxon word *lead* and its chemical symbol from the Latin word for the element, *plumbum*. The word *plumbing* is derived from this Latin word because of the Roman use of lead pipes for conducting water.

Lead is sometimes found as a pure metal in nature, but this form of the element is rather scarce. The most common ore of lead is galena, which consists of lead sulfide. The crushed ore, mixed with carbon, is roasted in a furnace and easily yields the metal.

Most of the lead produced in the United States is used to make the electrodes of lead storage batteries. The anode, or positive terminal, of the lead storage battery is made of a very porous form of lead called spongy lead. The cathode, or negative terminal, is made by packing a paste of lead oxide, called litharge, into a lead metal grid.

Lead is an important component of the solder used for making electrical connections on the circuit boards in computers and television sets. The glass screens of TV sets also contain lead to

Atomic Number **82**

Chemical Symbol **Pb**

Group **IVA**

Molten lead being formed into a musket ball.

shield the viewer from radiation. In fact, every TV set and computer contains about 0.5 pound of lead. The disposal of this lead poses a major environmental problem. The U.S. Environmental Protection Agency estimated that approximately 50,000 tons of lead from consumer electronic products were discarded in 1991. This figure represents a good share of the total amount of lead stored in municipal solid-waste disposal sites.

Many compounds of lead are used as paint pigments, because they are insoluble. Chrome yellow (lead chromate), red lead (lead oxide), and white lead (a basic lead carbonate) are all common pigments used to color paint. The use of lead in paint is now being restricted in many communities, however, because lead compounds have been found to be cumulative poisons. This means that the body retains the lead, and even small doses accumulated over a long period can be dangerous. Like many heavy metals, lead disables the enzymes in the body that are catalysts for essential biochemical reactions. In time, this can cause irreversible damage to the brain, liver, and kidneys. The lead poisoning of children living in tenements, caused by their eating flakes of lead-based paints, has been a particularly serious problem.

"Leaded" gasoline has also been recognized as a health hazard, and its use has been curtailed in the United States. Such gasoline contains lead tetraethyl as an additive to suppress engine "knocking," or detonation of the fuel–air mixture before it is fully compressed. Lead is also prohibited in the fuel for modern automobiles because it "poisons" the catalysts used in catalytic converters. In addition to this, lead shot used for hunting has become illegal in many areas of the country. This action was necessary because waterfowl such as ducks and geese injured by birdshot showed high concentrations of lead in their bodies and were thought unsafe to eat. Despite their possible hazards, however, lead water pipes continue to be used throughout the world.

Lead takes its chemical symbol from the Latin word for the element, plumbum. *The word* plumbing *is derived from this Latin word because of the Roman use of lead pipes for conducting water.*

Lead oxide is still used to produce a very heavy fine glass called crystal. This glass has a high index of refraction, which means that it can bend light rays at sharp angles. When cut in a way that gives it many faces and angles, crystal glass sparkles with a characteristic brilliance.

Among other uses, lead is mixed with other metals to make the metal used for printing type and is used to make the lead bricks commonly employed as a radiation shield around nuclear reactors and X-ray machines, because the dense metal absorbs radiation very effectively.

Lead has many isotopes. Some are end products of the decay of naturally occurring radioactive elements such as uranium and thorium. Because the half-life of uranium is many billions of years—the same order of magnitude as the age of the Earth—geologists can use the concentration of lead isotopes in uranium-bearing rocks to determine the age of those rocks.

IA																	VIIIA
H	IA											IIIA	IVA	VA	VIA	VIIA	He
Li	Be											B	C	N	O	F	Ne
Na	Mg	IIIB	IVB	VB	VIB	VIIB		VIIIB		IB	IIB	Al	Si	P	S	Cl	Ar
K	Ca	Sc	Ti	V	Cr	Mn	Fe	Co	Ni	Cu	Zn	Ga	Ge	As	Se	Br	Kr
Rb	Sr	Y	Zr	Nb	Mo	Tc	Ru	Rh	Pd	Ag	Cd	In	Sn	Sb	Te	I	Xe
Cs	Ba	*La	Hf	Ta	W	Re	Os	Ir	Pt	Au	Hg	Tl	Pb	Bi	Po	At	Rn
Fr	Ra	†Ac	Rf	Db	Sg	Bh	Hs	Mt	Ds	Rg	Cn	Uut	Uuq	Uup	Uuh	Uus	Uuo

*	Ce	Pr	Nd	Pm	Sm	Eu	Gd	Tb	Dy	Ho	Er	Tm	Yb	Lu
†	Th	Pa	U	Np	Pu	Am	Cm	Bk	Cf	Es	Fm	Md	No	Lr

Bismuth

Atomic Number **83**

Chemical Symbol **Bi**

Group **VA—Post-transition Metal**

Bismuth is a white, heavy, brittle metal that has a slight yellowish tinge. It is the last element in Group VA and is fairly resistant to corrosion. When heated in air, it burns with a blue flame to produce yellow clouds of its oxide.

Bismuth has been known since the early 15th century but was long confused with tin and lead. Although its discovery is not well documented, Claude Geoffroy the Younger, a French nobleman, is usually credited with its identification in 1753. It derives its name from the German *weisse masse*, which means "white mass."

Bismuth is sometimes found in its metallic state in nature but more often in its principal ores, bismite (bismuth oxide) and

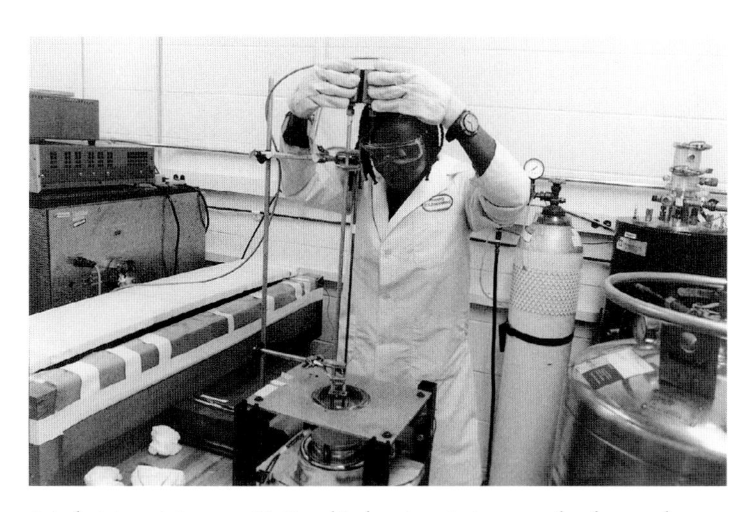

A technician at Argonne National Laboratory tests a sample of a powder that uses bismuth as a strength-enhancing agent.

Bismuth derives its name from the German weisse masse, *which means "white mass."*

bismuth glance (bismuth sulfide). It is often found in the company of copper, tin, and lead, and in the United States is usually recovered as a by-product of the refining of these metals.

Bismuth subnitrate has been used in medicine as an antacid in the treatment of ulcers. Bismuth oxide is a popular yellow pigment used in cosmetics.

Like water, bismuth is one of the few substances that expands when it changes from a liquid to a solid. This property is used to make alloys whose volume remains constant when they solidify. Metals alloyed with bismuth, for example, can be used for casts and molds that retain their exact dimensions even when filled with molten metals. Bismuth alloys with low melting points, such as Wood's meta—an alloy of bismuth, lead, tin, and cadmium—are used in fire alarms and sprinkler systems, and similar alloys are used as electrical fuses. Excessive electric current produces enough heat to melt these alloys, setting off fire alarms and sprinklers and breaking the circuit, through an electrical fuse, thus preventing damage to the electrical system.

IA																	VIIIA
H	IA											IIIA	IVA	VA	VIA	VIIA	He
Li	Be											B	C	N	O	F	Ne
Na	Mg	IIIB	IVB	VB	VIB	VIIB	VIIIB			IB	IIB	Al	Si	P	S	Cl	Ar
K	Ca	Sc	Ti	V	Cr	Mn	Fe	Co	Ni	Cu	Zn	Ga	Ge	As	Se	Br	Kr
Rb	Sr	Y	Zr	Nb	Mo	Tc	Ru	Rh	Pd	Ag	Cd	In	Sn	Sb	Te	I	Xe
Cs	Ba	*La	Hf	Ta	W	Re	Os	Ir	Pt	Au	Hg	Tl	Pb	Bi	Po	At	Rn
Fr	Ra	†Ac	Rf	Db	Sg	Bh	Hs	Mt	Ds	Rg	Cn	Uut	Uuq	Uup	Uuh	Uus	Uuo

*	Ce	Pr	Nd	Pm	Sm	Eu	Gd	Tb	Dy	Ho	Er	Tm	Yb	Lu
†	Th	Pa	U	Np	Pu	Am	Cm	Bk	Cf	Es	Fm	Md	No	Lr

Po

The discovery of polonium by Marie and Pierre Curie in 1898 defines one of the great moments in the history of science. Their discovery of this element led to the modern concept of the atomic nucleus and an understanding of its structure. The Curies were looking for the source of the radio-activity in the uranium-bearing mineral pitchblende. What puzzled them was that removing the uranium from this mineral still left radioactivity in the pitchblende. Their research finally led them to isolate two new elements, the first of which was polonium and the second radium.

Polonium was named for Poland, the native country of Marie Curie. It is sometimes referred to by its older name, radium F. It is a rare element, formed chiefly through the decay of radioactive uranium and thorium. It is found in uranium mines in concentrations as low as 100 millionths of a gram per ton of ore. Polonium has 27 known isotopes, more than any other element, and all of them are radioactive.

Marie Curie was awarded the Nobel Prize for chemistry in 1911 for the discovery of polonium and radium in 1898.

Polonium was named for Poland, the native country of its discoverer, Marie Curie.

The most common such isotope, and the one most readily available, is polonium-210. It has a half-life of only 138 days, making it approximately 5,000 times more radioactive than radium. Polonium-210 is quite dangerous, and great care has to be taken in handling even small amounts of this isotope.

Polonium-210 is a silvery metalloid that is quite volatile. It will almost completely vaporize at temperatures as low as 60°C. Its scientific and commercial uses are chiefly related to its radioactivity. As it decays, it emits alpha particles, each of which is essentially the nucleus of a helium atom, consisting of two protons and two neutrons. When an atom of polonium-210 ejects an alpha particle, it is changed, or "transmuted," into lead-206, a stable isotope of lead.

Small amounts of polonium-210 are occasionally used on dust-removal brushes to remove the static electric charge that forms on photographic film. In this application, the radiation emitted by the polonium ionizes the air through which it passes, and the resulting ions neutralize the electric charge on the film. In radiological laboratories, polonium-210 mixed with powdered beryllium is often used to produce large amounts of neutrons without the use of a nuclear reactor. The alpha particles emitted by the polonium enter the beryllium nucleus, initiating a nuclear reaction that causes the beryllium to emit neutrons. A special license issued by the U.S. Nuclear Regulatory Commission is required to operate this neutron source. Polonium is not only highly radioactive, but also extremely chemically toxic when taken into the body. The Nuclear Regulatory Commission has established a quantity of 6.8 million millionths of a gram as the maximum allowable amount of polonium-210 that can be ingested. This makes polonium-210 approximately 250,000 times more toxic than cyanide. An unprecedented example of the toxicity of polonium-210 made world headlines in 2006, when the former Russian spy, Alexander Litvinenko, was fatally poisoned by a small amount of polonium-210.

Astatine

IA	IA	IIIB	IVB	VB	VIB	VIIB	VIIIB	VIIIB	VIIIB	IB	IIB	IIIA	IVA	VA	VIA	VIIA	VIIIA
H																	He
Li	Be											B	C	N	O	F	Ne
Na	Mg											Al	Si	P	S	Cl	Ar
K	Ca	Sc	Ti	V	Cr	Mn	Fe	Co	Ni	Cu	Zn	Ga	Ge	As	Se	Br	Kr
Rb	Sr	Y	Zr	Nb	Mo	Tc	Ru	Rh	Pd	Ag	Cd	In	Sn	Sb	Te	I	Xe
Cs	Ba	*La	Hf	Ta	W	Re	Os	Ir	Pt	Au	Hg	Tl	Pb	Bi	Po	At	Rn
Fr	Ra	†Ac	Rf	Db	Sg	Bh	Hs	Mt	Ds	Rg	Cn	Uut	Uuq	Uup	Uuh	Uus	Uuo

*	Ce	Pr	Nd	Pm	Sm	Eu	Gd	Tb	Dy	Ho	Er	Tm	Yb	Lu
†	Th	Pa	U	Np	Pu	Am	Cm	Bk	Cf	Es	Fm	Md	No	Lr

At

All 20 of the known isotopes of astatine are radioactive, and the element is very short-lived. The isotope with the longest half-life is astatine-210, with a half-life of only 8.3 hours. Astatine was first made in 1940 by a team of radiochemists at the University of California at Berkeley. The principal members of the team were Dale R. Corsun, K. R. Mckenzie, and Emilio Segrè. They produced astatine by bombarding bismuth with alpha particles. They named the element from the Greek word *astatos*, which means "unstable."

Small quantities of astatine are produced naturally as the decay products of isotopes of uranium and thorium. The total amount of astatine estimated to be present in the Earth's crust at any one time is approximately 1 ounce. Only about 1 millionth of a gram of astatine has actually been produced artificially, and it is therefore not surprising that little is known about its properties. Astatine is a member of the halogen family, located just below iodine in Group VIIA of the periodic table, so that its chemistry should be fairly similar to that of iodine. There seems to be some evidence that astatine is slightly more metallic than iodine and that, like iodine, it is probably taken up by the thyroid gland.

Emilio Segrè was part of the team that created astatine in 1940.

Atomic Number **85**

Chemical Symbol **At**

Group **VIIA—The Halogens**

Radon

Atomic Number **86**

Chemical Symbol **Rn**

Group **VIIIA—The Noble Gases**

IA	IA	IIIB	IVB	VB	VIB	VIIB	VIIIB			IB	IIB	IIIA	IVA	VA	VIA	VIIA	VIIIA
H																	He
Li	Be											B	C	N	O	F	Ne
Na	Mg											Al	Si	P	S	Cl	Ar
K	Ca	Sc	Ti	V	Cr	Mn	Fe	Co	Ni	Cu	Zn	Ga	Ge	As	Se	Br	Kr
Rb	Sr	Y	Zr	Nb	Mo	Tc	Ru	Rh	Pd	Ag	Cd	In	Sn	Sb	Te	I	Xe
Cs	Ba	*La	Hf	Ta	W	Re	Os	Ir	Pt	Au	Hg	Tl	Pb	Bi	Po	At	Rn
Fr	Ra	†Ac	Rf	Db	Sg	Bh	Hs	Mt	Ds	Rg	Cn	Uut	Uuq	Uup	Uuh	Uus	Uuo

*	Ce	Pr	Nd	Pm	Sm	Eu	Gd	Tb	Dy	Ho	Er	Tm	Yb	Lu
†	Th	Pa	U	Np	Pu	Am	Cm	Bk	Cf	Es	Fm	Md	No	Lr

Rn

Radon is a gas with 20 known isotopes, all of which are radioactive. It is produced as one of the by-products of the radioactive decay of uranium and thorium, and it is the heaviest known gas, being about eight times heavier than air. As a noble gas, it is chemically unreactive.

Radon was discovered by the German physicist Friedrich Ernst Dorn in 1900 while studying the decay products of radium. He named it radium emanation because the gas seemed to come from the radium. William Ramsay and R. W. Whytlaw-Gray, two early investigators of the chemical properties of radon, later changed its name to niton, from the Latin word *nitens*, which means "shining." Since 1923, however, the element has been called radon.

Radon-222 is the longest lived isotope of radon, with a half-life of 3.82 days, and it is the isotope most generally available and studied. Radon-222 is found in substantial concentrations as a gas in the soil because trace amounts of uranium are present throughout the Earth's crust. The gas diffuses through the soil and into the air. The amount of radon in the air varies from one region to another, but it is considered a potential hazard in many homes in certain areas. The gas can diffuse into the home through basement floors and walls because pressure inside the house is always lower than outside. Without a constant interchange of fresh air from outside for air within the house, radon can build up to dangerous concentrations. This situation is particularly critical during the winter months when windows are likely to be kept shut. Inexpensive detectors are available to measure the amount of radon present in a home.

As early as the 16th century, it was known that uranium miners in Bohemia often died prematurely from diseases of the lung. We now know that the miners suffered from lung cancer caused by radon. As it decays, radon-222 emits alpha particles, which are essentially the nuclei of helium atoms, and simultaneously initiates a decay process that eventually produces lead-210. When radon-222 is inhaled, some of it will decay in the lungs before it can be exhaled. Exhaling removes much of the radon-222, but the decay product, lead-210, is also radioactive and settles in the lung. Its half-life of 20.4 years is much longer than that of radon-222, and it is not eliminated during breathing. It is this lead-210 that exposes the lungs to radiation for long periods and can produce lung cancer. Health officials in the United States have estimated that approximately 10 percent of all lung cancers are caused by radon.

Many homeowners, increasingly aware of the dangers posed by the presence of radon in homes, have installed radon detectors to warn them of potentially harmful levels of the gas.

Smoking a cigarette poses a radiation risk as well as a chemical one. While it is growing, tobacco is subject to contamination by radon from the soil, and the phosphate fertilizers used by planters are rich in uranium. As a result, the broad tobacco leaves become dusted with trace amounts of lead-210, and when this leaf is burned the inhaled smoke subjects the smoker to levels of radiation 1,000 times higher than those encountered by a worker in a nuclear power plant.

Despite the advances made in radiation therapy through the use of particle accelerators and isotopes such as cobalt-60, radon is still used in many hospitals for cancer therapy. It is usually pumped from a radium source and sealed into tiny glass vials called "seeds," which are implanted in patients at the sites of their tumors.

Francium

Atomic Number **87**

Chemical Symbol **Fr**

Group **IA—The Alkali Metals**

IA	IA	IIIB	IVB	VB	VIB	VIIB	VIIIB	VIIIB	VIIIB	IB	IIB	IIIA	IVA	VA	VIA	VIIA	VIIIA
H																	He
Li	Be											B	C	N	O	F	Ne
Na	Mg											Al	Si	P	S	Cl	Ar
K	Ca	Sc	Ti	V	Cr	Mn	Fe	Co	Ni	Cu	Zn	Ga	Ge	As	Se	Br	Kr
Rb	Sr	Y	Zr	Nb	Mo	Tc	Ru	Rh	Pd	Ag	Cd	In	Sn	Sb	Te	I	Xe
Cs	Ba	*La	Hf	Ta	W	Re	Os	Ir	Pt	Au	Hg	Tl	Pb	Bi	Po	At	Rn
Fr	Ra	†Ac	Rf	Db	Sg	Bh	Hs	Mt	Ds	Rg	Cn	Uut	Uuq	Uup	Uuh	Uus	Uuo

*	Ce	Pr	Nd	Pm	Sm	Eu	Gd	Tb	Dy	Ho	Er	Tm	Yb	Lu
†	Th	Pa	U	Np	Pu	Am	Cm	Bk	Cf	Es	Fm	Md	No	Lr

Fr

Francium is the heaviest of the alkali metals and one of the most unstable known. All of its isotopes are radioactive, yet even its longest lived isotope, francium-223, has a half-life of only 21 minutes. Of its 30 known isotopes, only francium-223 exists in nature. All of the other isotopes of francium are produced artificially in accelerators and nuclear reactors and are too unstable to be studied in any depth. The element was discovered in 1939 by Marguerite Perey, working at the Curie Institute in Paris. It is named for France, the country in which it was discovered.

Francium is produced by the radioactive decay of the elements uranium and thorium. It has been estimated that because of its short half-life, the Earth's crust contains less than 1 ounce of francium.

IA	IA	IIIB	IVB	VB	VIB	VIIB	VIIIB			IB	IIB	IIIA	IVA	VA	VIA	VIIA	VIIIA
H																	He
Li	Be											B	C	N	O	F	Ne
Na	Mg											Al	Si	P	S	Cl	Ar
K	Ca	Sc	Ti	V	Cr	Mn	Fe	Co	Ni	Cu	Zn	Ga	Ge	As	Se	Br	Kr
Rb	Sr	Y	Zr	Nb	Mo	Tc	Ru	Rh	Pd	Ag	Cd	In	Sn	Sb	Te	I	Xe
Cs	Ba	*La	Hf	Ta	W	Re	Os	Ir	Pt	Au	Hg	Tl	Pb	Bi	Po	At	Rn
Fr	Ra	†Ac	Rf	Db	Sg	Bh	Hs	Mt	Ds	Rg	Cn	Uut	Uuq	Uup	Uuh	Uus	Uuo

*	Ce	Pr	Nd	Pm	Sm	Eu	Gd	Tb	Dy	Ho	Er	Tm	Yb	Lu
†	Th	Pa	U	Np	Pu	Am	Cm	Bk	Cf	Es	Fm	Md	No	Lr

Radium

Radium was discovered in 1898 by the Polish-born French chemist Marie Sklodowska Curie and her husband, Pierre Curie. Along with the discovery of the electron and Einstein's theory of relativity, the discovery of radium marked the beginning of the modern era in science.

In their investigation of the radioactive properties of a uranium-bearing mineral called pitchblende, the Curies discovered that removing the uranium from the pitchblende still left it radioactive. Working with tons of the ore, they finally separated out a radioactive mixture that seemed to consist chiefly of barium. The barium was used as a precipitating reagent for polonium and other elements in the ore. When the mixture was heated in a flame to produce spectral lines, the Curies saw a totally unexpected, beautiful red color in addition to the colors normally associated with barium. The newly discovered spectral line corresponded to a new element. The Curies named it radium from the Latin word *radius*, meaning "ray." They also discovered the element polonium in the same way.

It took the Curies 4 more years to obtain a pure sample of radium,

Marie Curie, with her husband, Pierre, in their laboratory.

Atomic Number **88**

Chemical Symbol **Ra**

Group **IIA—The Alkaline-Earth Metals**

Radium was discovered in 1898 by Marie Curie and her husband, Pierre Curie. They named it radium from the Latin word radius, *meaning "ray."*

which they isolated from a solution of radium chloride by electrolysis. In 1911, Marie Curie was awarded the Nobel Prize in chemistry for her discovery of these elements. It was her second Nobel Prize; she had shared the first with her husband and the French scientist Henri Becquerel in 1903 for the discovery of radioactivity.

Radium is the last member of the alkaline-earth elements, and like them it is a metal. The pure metal has a brilliant white color and is so luminescent that it glows in the dark, giving off a faint blue color. Before the dangers of radioactivity were understood, radium was used to make luminous paints for the dials of watches and clocks so that they could be read in the dark. The "rays" emanating from radium probably served as the reason for choosing the Latin equivalent of "ray" for its name. The element darkens when it is exposed to air, forming a compound with nitrogen called a nitride. It also reacts with water, decomposing it and forming radium hydroxide. In research or medicine, radium is usually used in the form of a salt such as radium chloride.

Radium is present in all uranium-bearing ores because it is one of the decay products of uranium. About 1 gram of radium is present in every 7 tons of pitchblende, deposits of which are found in the Czech Republic and Slovakia and parts of Africa. Deposits of uranium ore are also found in the United States, chiefly in Utah and New Mexico. The total world production of radium amounts to not much more than 5 pounds annually.

Radium has 25 known isotopes. The most common, and the one discovered by the Curies, is radium-226. It has a half-life of 1,630 years. The international unit of measurement of all radioactive substances is called the curie, in honor of the Curies, and uses radium-226 as its standard. One curie of radiation is defined as the radioactivity of 1 gram of radium-226. This corresponds to 37 billion radium nuclei disintegrating in 1 second. Most radioactive materials used in schools for demonstrations have much lower levels of activity, generally in the microcurie range. (The prefix *micro* means 1 millionth.)

Radium is used in many medical facilities to generate the radioactive gas radon, which is used for cancer therapy.

IA																	VIIIA
H	IA											IIIA	IVA	VA	VIA	VIIA	He
Li	Be											B	C	N	O	F	Ne
Na	Mg	IIIB	IVB	VB	VIB	VIIB		VIIIB		IB	IIB	Al	Si	P	S	Cl	Ar
K	Ca	Sc	Ti	V	Cr	Mn	Fe	Co	Ni	Cu	Zn	Ga	Ge	As	Se	Br	Kr
Rb	Sr	Y	Zr	Nb	Mo	Tc	Ru	Rh	Pd	Ag	Cd	In	Sn	Sb	Te	I	Xe
Cs	Ba	*La	Hf	Ta	W	Re	Os	Ir	Pt	Au	Hg	Tl	Pb	Bi	Po	At	Rn
Fr	Ra	†Ac	Rf	Db	Sg	Bh	Hs	Mt	Ds	Rg	Cn	Uut	Uuq	Uup	Uuh	Uus	Uuo

*	Ce	Pr	Nd	Pm	Sm	Eu	Gd	Tb	Dy	Ho	Er	Tm	Yb	Lu
†	Th	Pa	U	Np	Pu	Am	Cm	Bk	Cf	Es	Fm	Md	No	Lr

Actinium

Ac

Actinium is a radioactive element produced naturally by the radioactive decay of the long-lived elements uranium and thorium. Very small amounts of it have been produced artificially, and it has very limited commercial or scientific application.

Actinium is a metal, and like radium it glows in the dark. Its name is taken from the Greek word *aktinos*, which means "ray" or "beam." It was first discovered by the French scientist André Debierne in 1899 and then rediscovered independently by the German chemist Friedrich Otto Giesel in 1902.

There are 26 known isotopes of actinium and all are radioactive. The most important is actinium-227, a product of the decay of uranium-235. It has a half-life of 21.6 years. Its chemical properties resemble those of the rare earth element lanthanum. Actinium also resembles lanthanum in that it is the first element in the series of elements called the actinides, which are analogous to the lanthanides. The actinide series, sometimes called the Second Inner Transition Series, contains the elements from actinium through lawrencium (atomic number 103). Like the rare earths, these elements add electrons to an inner orbital shell, the second such shell from the outer, valence shell, and consequently have similar chemical and physical properties.

Actinide recycling at Argonne National Laboratory.

Atomic Number 89

Chemical Symbol Ac

Group IIIB—Transition Element (The Actinides)

Thorium

IA	IA	IIIB	IVB	VB	VIB	VIIB	VIIIB			IB	IIB	IIIA	IVA	VA	VIA	VIIA	VIIIA
H																	He
Li	Be											B	C	N	O	F	Ne
Na	Mg											Al	Si	P	S	Cl	Ar
K	Ca	Sc	Ti	V	Cr	Mn	Fe	Co	Ni	Cu	Zn	Ga	Ge	As	Se	Br	Kr
Rb	Sr	Y	Zr	Nb	Mo	Tc	Ru	Rh	Pd	Ag	Cd	In	Sn	Sb	Te	I	Xe
Cs	Ba	*La	Hf	Ta	W	Re	Os	Ir	Pt	Au	Hg	Tl	Pb	Bi	Po	At	Rn
Fr	Ra	†Ac	Rf	Db	Sg	Bh	Hs	Mt	Ds	Rg	Cn	Uut	Uuq	Uup	Uuh	Uus	Uuo

*	Ce	Pr	Nd	Pm	Sm	Eu	Gd	Tb	Dy	Ho	Er	Tm	Yb	Lu
†	Th	Pa	U	Np	Pu	Am	Cm	Bk	Cf	Es	Fm	Md	No	Lr

Atomic Number 90

Chemical Symbol Th

Group IIIB—Transition Element (The Actinides)

Th

Thorium is a radioactive, silvery-white metal that tarnishes very slowly when exposed to air. After a few months, it reacts with the air to form a black oxide. Pure thorium is very soft and malleable. Its oxide has a melting point of 3,300°C, one of the highest of any oxide, and when finely subdivided can be ignited to burn in air, producing a brilliant white light.

Thorium was discovered in 1828 by the Swedish chemist Jöns Jakob Berzelius, who named it for Thor, the Scandinavian god of war. He was unaware of its radioactivity because that was not known as a physical process until it was discovered by Henri Becquerel and the Curies in 1897. Thorium-232, the isotope of thorium that occurs naturally, is actually very weakly radioactive. Its half-life is an enormous 14 billion years, so that very little of it decays in a short period. The radiation it emits can nevertheless fog photographic film if left in contact with the film for several hours.

Thorium-containing ores are about as abundant as those of lead in the Earth's crust and about three or four times more abundant than uranium ores. Monazite sand, some of which is found as beach sand in Florida, can contain up to 10 percent thorium. This sand is used for the commercial preparation of the element.

Thorium shows great promise of becoming an important source of nuclear energy in the future. When thorium-232 is subjected to a beam of neutrons, it undergoes several nuclear transformations to form an isotope of uranium called uranium-233. Uranium-233 can undergo nuclear fission in the same manner as uranium-235, the isotope now used throughout the world

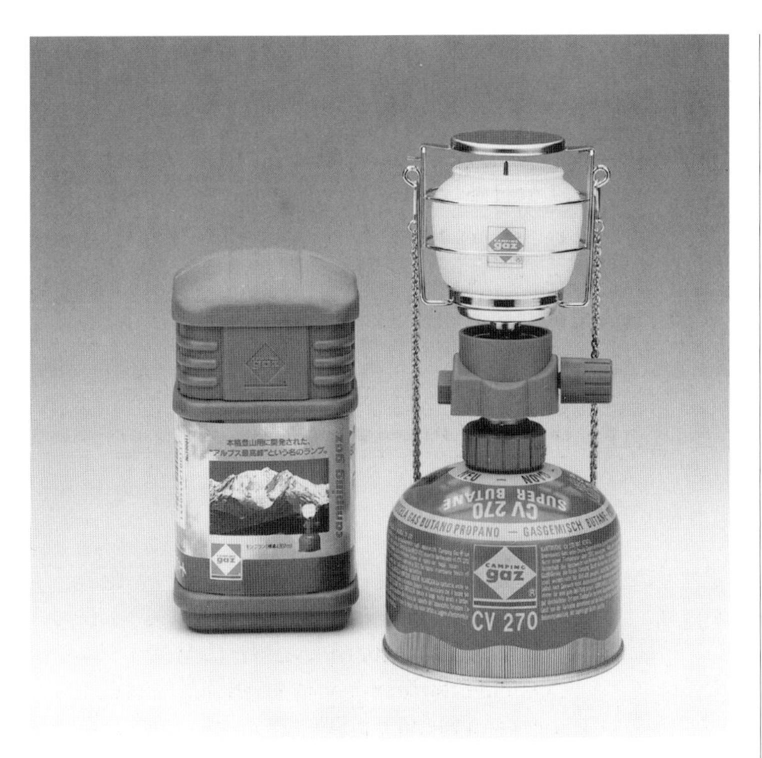

Thorium oxide is used to make the mantles of portable gas lamps.

as a commercial source of nuclear power. Several experimental prototype reactors using uranium-233 are currently under construction. Given the large quantities of thorium available on Earth, there are hopes that it will become a major future source of energy.

Thorium has some 25 known isotopes, with half-lives ranging from one-tenth of a millionth of a second to the 14 billion years of thorium-232, the longest lived isotope of the element. When thorium-232 decays, it transmutes itself into 11 different elements before ending up as lead-208, a stable isotope of lead. This series of transformations is called the thorium decay series.

Despite its radioactivity, thorium metal and its compounds have several commercial applications. The metal serves as an alloying element for magnesium that must be subjected to high temperatures. Thorium also serves as a very efficient emitter of electrons for electronic devices. The brilliant light that thorium oxide emits when burning makes it useful in fabricating certain types of portable gas lamps.

Thorium was discovered in 1828 by the Swedish chemist Jöns Jakob Berzelius, who named it for Thor, the Scandinavian god of war.

Protactinium

IA																	VIIIA
H	IA											IIIA	IVA	VA	VIA	VIIA	He
Li	Be											B	C	N	O	F	Ne
Na	Mg	IIIB	IVB	VB	VIB	VIIB		VIIIB		IB	IIB	Al	Si	P	S	Cl	Ar
K	Ca	Sc	Ti	V	Cr	Mn	Fe	Co	Ni	Cu	Zn	Ga	Ge	As	Se	Br	Kr
Rb	Sr	Y	Zr	Nb	Mo	Tc	Ru	Rh	Pd	Ag	Cd	In	Sn	Sb	Te	I	Xe
Cs	Ba	*La	Hf	Ta	W	Re	Os	Ir	Pt	Au	Hg	Tl	Pb	Bi	Po	At	Rn
Fr	Ra	†Ac	Rf	Db	Sg	Bh	Hs	Mt	Ds	Rg	Cn	Uut	Uuq	Uup	Uuh	Uus	Uuo

*	Ce	Pr	Nd	Pm	Sm	Eu	Gd	Tb	Dy	Ho	Er	Tm	Yb	Lu
†	Th	Pa	U	Np	Pu	Am	Cm	Bk	Cf	Es	Fm	Md	No	Lr

Pa

Protactinium is one of the scarcest and most expensive of the naturally existing elements. It is radioactive and is found in such uranium ores as pitchblende, where it is produced as one of the decay products of uranium.

Only a few hundred grams of protactinium are available for study. This meager amount was largely produced in England some 30 years ago, where it was extracted from 60 tons of ore at a cost of half a million dollars.

Protactinium was discovered by the German physicists Kasimir Fajans and O. H. Gohring in 1913 during an investigation of the elements produced in the decay of uranium. The two scientists actually named their new element brevium, but the name was later changed to protactinium in 1949. Its name is derived from the Greek word *protos*, meaning "first," and *actinium*, because the decay product of protactinium is actinium. There are some 22 known isotopes of protactinium, of which the most important is protactinium-231, with a half-life of 32,500 years.

Not much is known about the chemical and physical properties of protactinium. The pure metal was finally isolated in 1934, and it is a silvery-white metal with a bright luster that it loses very slowly in air, through oxidation. Protactinium is extremely toxic and must be handled with great care.

Atomic Number 91

Chemical Symbol Pa

Group IIIB—Transition Element (The Actinides)

IA	IA	IIIB	IVB	VB	VIB	VIIB	VIIIB			IB	IIB	IIIA	IVA	VA	VIA	VIIA	VIIIA
H																	He
Li	Be											B	C	N	O	F	Ne
Na	Mg											Al	Si	P	S	Cl	Ar
K	Ca	Sc	Ti	V	Cr	Mn	Fe	Co	Ni	Cu	Zn	Ga	Ge	As	Se	Br	Kr
Rb	Sr	Y	Zr	Nb	Mo	Tc	Ru	Rh	Pd	Ag	Cd	In	Sn	Sb	Te	I	Xe
Cs	Ba	*La	Hf	Ta	W	Re	Os	Ir	Pt	Au	Hg	Tl	Pb	Bi	Po	At	Rn
Fr	Ra	†Ac	Rf	Db	Sg	Bh	Hs	Mt	Ds	Rg	Cn	Uut	Uuq	Uup	Uuh	Uus	Uuo

*	Ce	Pr	Nd	Pm	Sm	Eu	Gd	Tb	Dy	Ho	Er	Tm	Yb	Lu
†	Th	Pa	U	Np	Pu	Am	Cm	Bk	Cf	Es	Fm	Md	No	Lr

Uranium

Atomic Number **92**

Chemical Symbol **U**

Group IIIB—Transition Element (The Actinides)

Uranium is the last and heaviest of the natural elements. Most people associate it with nuclear reactors and the atomic bomb as a source of enormous energy. Once considered scarce, it is now found in many minerals, including pitchblende, uranite, "yellowcake," and monazite sands. In the United States, the ownership and sale of uranium are strictly controlled by the U.S. Nuclear Regulatory Commission.

The use of uranium compounds to color glass and ceramic glazes dates back thousands of years. The German chemist Martin Klaproth, in 1789, was the first investigator to realize that pitchblende contained an unknown element. It was not until 1841 that the French chemist Eugène-Melchior Péligot first isolated and identified uranium. It was named for the planet Uranus. In 1896, the French physicist Henri Becquerel discovered that uranium was radioactive. It was the first radioactive element to be discovered.

Naturally occurring uranium is a dense, silvery-white metal that quickly acquires a dark oxide coating when exposed to air. A typical sample of uranium essentially consists of two isotopes, uranium-238 (99.2798 percent) and uranium-235 (0.7171 percent), with a trace of a third isotope, uranium-234 (0.0031 percent). It is not surprising that uranium-238 is the dominant isotope of the element; its half-life of 4.6 billion years makes it the longest lived of the three isotopes. A long half-life means that an isotope is less active and that fewer of its atomic nuclei disintegrate in any given period. Uranium-235 has a half-life of 700 million years, whereas uranium-234 has a half-life of only 25 million years.

Enrico Fermi's work with uranium led to the first self-sustaining nuclear chain reaction in 1944. Fermi went on to work on the development of the atomic bomb.

Given the long half-lives of these three isotopes, uranium itself is only weakly radioactive. But in the process of decaying, uranium does create many new highly radioactive isotopes, such as those of radon and polonium, before finally reaching a stable state as an isotope of lead. This chain of isotopes is called the uranium decay series.

In the late 1930s two German scientists, Lise Meitner and Otto Hahn, discovered that bombarding uranium with neutrons produced such elements as barium and krypton. The atoms of these elements were approximately half the size of the uranium atom. Where were they coming from? It was Lise Meitner and her nephew, the German physicist Otto R. Frisch, who first visualized the result as a fracture of the uranium nucleus into two fragments of intermediate size. They called the process, initiated by the uranium nucleus observing a neutron, nuclear fission. A large amount of energy, as well as several additional neutrons, was released in the process. Subsequent research quickly demonstrated that it was the uranium-235 isotope that was fissioning.

The ability of the neutrons released during the fission of the uranium nucleus to themselves split other uranium nuclei was quickly utilized by scientists to create a self-sustaining chain reaction. When controlled, this reaction produces the energy we obtain from nuclear reactors. When uncontrolled, it can produce an atomic explosion. The first self-sustaining nuclear chain reaction was achieved by the Italian American scientist Enrico Fermi at the University of Chicago in 1944.

Uranium is the last and heaviest of the natural elements.

The uranium fuel used for nuclear reactors is often enriched by increasing its percentage of uranium-235. The most common method for doing this uses the diffusion of gaseous uranium hexafluoride through thousands of porous membranes to obtain a fuel with the desired uranium-235 content. In the process of diffusion, which is the name given to the spontaneous movement of gases from regions of higher concentration to those of lower concentration, heavier gases move more slowly than light ones. Because uranium-238 is slightly heavier than uranium-235, it is eventually "left behind" as uranium-235 atoms accumulate at the end of the diffusion process.

When uranium-238 absorbs neutrons, it undergoes a number of reactions that transform it into plutonium-239, an isotope whose ability to fission is similar to uranium-235. This ability to fission is made use of in the so-called breeder reactors where new fuel in the form of plutonium-239 is produced as uranium-235, the original fuel powering the reactor is used up. Plutonium-239 is also used in the manufacture of atomic bombs.

After the creation of the atomic bomb, the public was captivated by the potential of nuclear energy. Here, a woman experiments with her very own sample of uranium.

Depleted uranium is uranium from which most of the uranium-235 atoms have decayed or been removed. It is used to produce armor-piercing antitank shells, ballast for missile reentry systems, glazes for ceramics, and shielding against radiation. Great care must be used in handling uranium-containing materials because, in addition to posing a radiation hazard, uranium and its compounds are highly toxic.

The age of rocks containing uranium can be dated by measuring the ratio of the remaining uranium-238 to the amount of lead-206, the last element in the uranium-238 decay series. This dating method has shown that the oldest rocks found on Earth are approximately 4.5 billion years old.

Neptunium

IA																	VIIIA
H	IA											IIIA	IVA	VA	VIA	VIIA	He
Li	Be											B	C	N	O	F	Ne
Na	Mg	IIIB	IVB	VB	VIB	VIIB		VIIIB		IB	IIB	Al	Si	P	S	Cl	Ar
K	Ca	Sc	Ti	V	Cr	Mn	Fe	Co	Ni	Cu	Zn	Ga	Ge	As	Se	Br	Kr
Rb	Sr	Y	Zr	Nb	Mo	Tc	Ru	Rh	Pd	Ag	Cd	In	Sn	Sb	Te	I	Xe
Cs	Ba	*La	Hf	Ta	W	Re	Os	Ir	Pt	Au	Hg	Tl	Pb	Bi	Po	At	Rn
Fr	Ra	†Ac	Rf	Db	Sg	Bh	Hs	Mt	Ds	Rg	Cn	Uut	Uuq	Uup	Uuh	Uus	Uuo

*	Ce	Pr	Nd	Pm	Sm	Eu	Gd	Tb	Dy	Ho	Er	Tm	Yb	Lu
†	Th	Pa	U	**Np**	Pu	Am	Cm	Bk	Cf	Es	Fm	Md	No	Lr

Np

Neptunium was the first artificially produced transuranium element. The prefix *trans* indicates that an element is "beyond" uranium in the periodic table. Neptunium derives its name from the planet Neptune because that is the planet beyond Uranus (the planet that gave its name to uranium) in our solar system.

Neptunium was first produced by the American physicists Edwin M. McMillan and Philip H. Abelson in 1940. Working at the cyclotron at the University of California at Berkeley, they produced neptunium by bombarding uranium with neutrons. It is now known that trace quantities of neptunium do actually exist in nature as a result of the action of neutrons present in uranium ore.

Currently, 18 isotopes of neptunium have been produced, all of them radioactive. The most important, and the first to be produced, was neptunium-237, with a half-life of 2.1 million years.

Neptunium is currently being made in nuclear reactors as a by-product of the generation of plutonium. It is a fairly reactive, silvery metal.

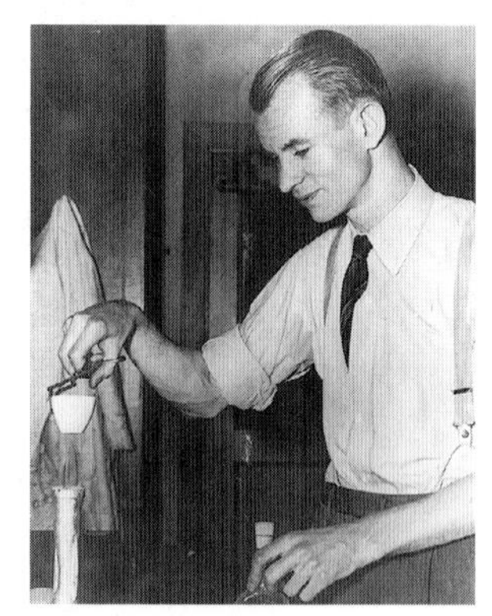

Dr. Edwin McMillan, one of the discoverers of neptunium, at work in his laboratory at the University of California at Berkeley.

Atomic Number 93

Chemical Symbol Np

Group IIIB—Transition Element (The Actinides)

IA																	VIIIA
H	IA											IIIA	IVA	VA	VIA	VIIA	He
Li	Be											B	C	N	O	F	Ne
Na	Mg	IIIB	IVB	VB	VIB	VIIB		VIIIB		IB	IIB	Al	Si	P	S	Cl	Ar
K	Ca	Sc	Ti	V	Cr	Mn	Fe	Co	Ni	Cu	Zn	Ga	Ge	As	Se	Br	Kr
Rb	Sr	Y	Zr	Nb	Mo	Tc	Ru	Rh	Pd	Ag	Cd	In	Sn	Sb	Te	I	Xe
Cs	Ba	*La	Hf	Ta	W	Re	Os	Ir	Pt	Au	Hg	Tl	Pb	Bi	Po	At	Rn
Fr	Ra	†Ac	Rf	Db	Sg	Bh	Hs	Mt	Ds	Rg	Cn	Uut	Uuq	Uup	Uuh	Uus	Uuo

*	Ce	Pr	Nd	Pm	Sm	Eu	Gd	Tb	Dy	Ho	Er	Tm	Yb	Lu
†	Th	Pa	U	Np	Pu	Am	Cm	Bk	Cf	Es	Fm	Md	No	Lr

Plutonium

Atomic Number **94**

Chemical Symbol **Pu**

Group IIIB—Transition Element (The Actinides)

Pu

Plutonium is the most important of the transuranium elements, all of which follow uranium in the periodic table and all of which are artificially made. Plutonium was the second transuranium element to be synthesized (neptunium was the first). In 1941, the celebrated American chemist Glenn T. Seaborg used the cyclotron at the University of California at Berkeley to discover

Plutonium was used to power electronic equipment on this Apollo lunar excursion module (LEM) in 1971.

plutonium. The element was produced by bombarding uranium with deuterons. Deuterons are the nuclei of the hydrogen isotope deuterium and contain a neutron as well as a proton. Plutonium was named for Pluto, the planet that follows Neptune (which gave its name to neptunium) in our solar system. Because the United States was at war with Germany and Japan during this period, the discovery of plutonium was not announced until 1946 for security reasons.

Plutonium is a fairly active, silvery metal that tarnishes in air to form an oxide with a slightly yellow color. A sample of the metal feels warm to the touch because of the energy released by its own radiation. The element is highly toxic, and special care is required to handle it safely.

In 1941, American chemist Glenn T. Seaborg used the cyclotron at the University of California at Berkeley to discover plutonium.

Plutonium has 15 known isotopes, and all are radioactive. Plutonium-239, with a half-life of 24,400 years, is the most important because it readily fissions when bombarded by thermal neutrons. Like uranium-235, the nuclei of its atoms split into two intermediate-size nuclei (called fission fragments), releasing large amounts of energy and producing more neutrons to sustain a chain reaction. Mixed with powdered beryllium, it is an effective source of neutrons for scientific work in laboratories and universities. Here the alpha particles emitted by the plutonium interact with beryllium nuclei to produce the neutrons.

Plutonium can be produced in huge quantities of tens of thousands of kilograms per year in nuclear reactors. Special "breeder" reactors have the sole function of producing plutonium, but it is also produced in ordinary nuclear reactors. The plutonium generated in the latter way is fairly easy to remove from the uranium by chemical separation. Huge diffusion plants, like the ones needed to separate uranium-235 from uranium-238, are not required. The abundance of plutonium has made it the material of choice for nuclear weapons.

Plutonium-238, with a half-life of 87 years, emits only alpha particles (helium nuclei), which are easily stopped by shielding. Although it does not fission, the energy of the alpha particles it gives off can be converted into heat, and this heat can then be converted into electricity by a thermoelectric device. The yield is about one-half a watt of power for every gram of plutonium. This has made it useful for power sources that are remote, such as on the Apollo lunar excursion module, or difficult to get at, such as a pacemaker for the heart.

IA	IA	IIIB	IVB	VB	VIB	VIIB	VIIIB			IB	IIB	IIIA	IVA	VA	VIA	VIIA	VIIIA
H																	He
Li	Be											B	C	N	O	F	Ne
Na	Mg											Al	Si	P	S	Cl	Ar
K	Ca	Sc	Ti	V	Cr	Mn	Fe	Co	Ni	Cu	Zn	Ga	Ge	As	Se	Br	Kr
Rb	Sr	Y	Zr	Nb	Mo	Tc	Ru	Rh	Pd	Ag	Cd	In	Sn	Sb	Te	I	Xe
Cs	Ba	*La	Hf	Ta	W	Re	Os	Ir	Pt	Au	Hg	Tl	Pb	Bi	Po	At	Rn
Fr	Ra	†Ac	Rf	Db	Sg	Bh	Hs	Mt	Ds	Rg	Cn	Uut	Uuq	Uup	Uuh	Uus	Uuo

*	Ce	Pr	Nd	Pm	Sm	Eu	Gd	Tb	Dy	Ho	Er	Tm	Yb	Lu
†	Th	Pa	U	Np	Pu	Am	Cm	Bk	Cf	Es	Fm	Md	No	Lr

Americium is a transuranium element, one of the elements that follows uranium in the periodic table. Like all of the transuranium elements, it is artificially made. It was the fourth transuranium element to be discovered and was named after the place where it was discovered, America.

Americium was discovered in 1944 by a team of chemists working under the leadership of the American chemist Glenn T. Seaborg. This took place during World War II, and the work was done at the former Metallurgical Laboratory at the University of Chicago, a laboratory that is today the world-famous Argonne National Laboratory. Seaborg's team produced americium-241, one of the 14 known isotopes of americium, all of which are radioactive, by bombarding plutonium with neutrons. Americium-241 emits both alpha particles (helium nuclei) and gamma rays (radiation similar to X rays) and has a half-life of 470 years.

Americium-241 is made in large quantities in nuclear reactors. The intense gamma radiation it emits has made it useful as a portable source of X rays. It is also used in home smoke detectors, in which alpha particles emitted by this isotope ionize the surrounding air by stripping electrons from gas molecules. Ionized air is a fairly good conductor of electricity; smoke particles in the air reduce its electrical conductivity and generate a signal that triggers the alarm.

A team led by American chemist Glenn T. Seaborg discovered americium in 1944.

Atomic Number **95**

Chemical Symbol **Am**

Group IIIB—Transition Element (The Actinides)

Curium

IA	IA	IIIB	IVB	VB	VIB	VIIB	VIIIB	VIIIB	VIIIB	IB	IIB	IIIA	IVA	VA	VIA	VIIA	VIIIA
H																	**He**
Li	**Be**											**B**	**C**	**N**	**O**	**F**	**Ne**
Na	**Mg**											**Al**	**Si**	**P**	**S**	**Cl**	**Ar**
K	**Ca**	**Sc**	**Ti**	**V**	**Cr**	**Mn**	**Fe**	**Co**	**Ni**	**Cu**	**Zn**	**Ga**	**Ge**	**As**	**Se**	**Br**	**Kr**
Rb	**Sr**	**Y**	**Zr**	**Nb**	**Mo**	**Tc**	**Ru**	**Rh**	**Pd**	**Ag**	**Cd**	**In**	**Sn**	**Sb**	**Te**	**I**	**Xe**
Cs	**Ba**	***La**	**Hf**	**Ta**	**W**	**Re**	**Os**	**Ir**	**Pt**	**Au**	**Hg**	**Tl**	**Pb**	**Bi**	**Po**	**At**	**Rn**
Fr	**Ra**	**†Ac**	**Rf**	**Db**	**Sg**	**Bh**	**Hs**	**Mt**	**Ds**	**Rg**	**Cn**	**Uut**	**Uuq**	**Uup**	**Uuh**	**Uus**	**Uuo**

*	**Ce**	**Pr**	**Nd**	**Pm**	**Sm**	**Eu**	**Gd**	**Tb**	**Dy**	**Ho**	**Er**	**Tm**	**Yb**	**Lu**
†	**Th**	**Pa**	**U**	**Np**	**Pu**	**Am**	**Cm**	**Bk**	**Cf**	**Es**	**Fm**	**Md**	**No**	**Lr**

Cm

Curium is a transuranium element, one of the elements that follow uranium in the periodic table. Like all of the transuranium elements, it is artificially made. It was discovered in 1944 by Glenn T. Seaborg, Ralph A. James, and Albert Ghiorso, who produced it by using a cyclotron at the University of California at Berkeley to bombard plutonium-239 with alpha particles (helium nuclei). The chemical identification of curium took place during World War II at the former Metallurgical Laboratory of the University of Chicago, known today as the Argonne National Laboratory. Seaborg, James, and Ghiorso named their new element in honor of Pierre and Marie Curie, the famous couple who discovered radium and polonium.

Curium is a silvery-white metal that is fairly reactive. Oxide and curium fluoride compounds of the element have been prepared and studied. Curium has 14 known isotopes, all of which are radioactive. The first to be discovered, and the third transuranium element to be made, was curium-242, with a half-life of only 163 days. The most stable of the curium isotopes is curium-247 with a half-life of 16 million years. A great deal of interest has been shown in the possibility of using curium-242 and curium-244 as sources of energy in remote areas. Fairly large quantities of curium-244, with a half-life of 18 years, can be made by subjecting plutonium to neutron bombardment in a nuclear reactor. The radiation these isotopes emit can be easily converted into heat and then into electricity by thermoelectric devices. Although it has a relatively short half-life, the power output of curium-242 is an impressive 2 to 3 watts per gram. These compact units are useful for pacemakers, remote navigational buoys, and space missions.

Atomic Number **96**

Chemical Symbol **Cm**

Group IIIB—Transition Element (The Actinides)

IA																	VIIIA
H	IA											IIIA	IVA	VA	VIA	VIIA	He
Li	Be											B	C	N	O	F	Ne
Na	Mg	IIIB	IVB	VB	VIB	VIIB	VIIIB			IB	IIB	Al	Si	P	S	Cl	Ar
K	Ca	Sc	Ti	V	Cr	Mn	Fe	Co	Ni	Cu	Zn	Ga	Ge	As	Se	Br	Kr
Rb	Sr	Y	Zr	Nb	Mo	Tc	Ru	Rh	Pd	Ag	Cd	In	Sn	Sb	Te	I	Xe
Cs	Ba	*La	Hf	Ta	W	Re	Os	Ir	Pt	Au	Hg	Tl	Pb	Bi	Po	At	Rn
Fr	Ra	†Ac	Rf	Db	Sg	Bh	Hs	Mt	Ds	Rg	Cn	Uut	Uuq	Uup	Uuh	Uus	Uuo

*	Ce	Pr	Nd	Pm	Sm	Eu	Gd	Tb	Dy	Ho	Er	Tm	Yb	Lu
†	Th	Pa	U	Np	Pu	Am	Cm	Bk	Cf	Es	Fm	Md	No	Lr

Berkelium, like so many of the actinium transition elements, was discovered in Berkeley, California. The discoverers of the element, the team consisting of Glenn T. Seaborg, Stanley Thompson, and Albert Ghiorso, named it in honor of the city. They synthesized it in 1949 using a cyclotron to bombard a sample of americium-241 with alpha particles (helium nuclei). Berkelium-243, the isotope they produced, was the fifth actinide to be created in a laboratory. It has a half-life of only 4.6 hours, hardly long enough to study in any detail.

Fourteen more isotopes of berkelium have been synthesized, and all are radioactive. The longest-lived is berkelium-249, with a half-life of 314 days, which was made by subjecting curium-244 to intense neutron irradiation. Using berkelium-249, it was possible in 1962 to produce 3 billionths of a gram of berkelium chloride, but the pure metal has never been isolated. Given its scarcity, it is not surprising that no commercial or scientific applications for berkelium have been developed.

In 1949, Albert Ghiorso (above) and two of his colleagues produced berkelium using the cyclotron at the University of California at Berkeley.

Californium

Atomic Number 98

Chemical Symbol Cf

Group IIIB—Transition Element (The Actinides)

IA																	VIIIA
H	IA											IIIA	IVA	VA	VIA	VIIA	He
Li	Be											B	C	N	O	F	Ne
Na	Mg	IIIB	IVB	VB	VIB	VIIB	VIIIB			IB	IIB	Al	Si	P	S	Cl	Ar
K	Ca	Sc	Ti	V	Cr	Mn	Fe	Co	Ni	Cu	Zn	Ga	Ge	As	Se	Br	Kr
Rb	Sr	Y	Zr	Nb	Mo	Tc	Ru	Rh	Pd	Ag	Cd	In	Sn	Sb	Te	I	Xe
Cs	Ba	*La	Hf	Ta	W	Re	Os	Ir	Pt	Au	Hg	Tl	Pb	Bi	Po	At	Rn
Fr	Ra	†Ac	Rf	Db	Sg	Bh	Hs	Mt	Ds	Rg	Cn	Uut	Uuq	Uup	Uuh	Uus	Uuo

*	Ce	Pr	Nd	Pm	Sm	Eu	Gd	Tb	Dy	Ho	Er	Tm	Yb	Lu
†	Th	Pa	U	Np	Pu	Am	Cm	Bk	Cf	Es	Fm	Md	No	Lr

Cf

Californium was discovered in 1950 by the same team that discovered berkelium, but with one new member. In 1950, Stanley Thompson, Kenneth Street, Jr., Albert Ghiorso, and Glenn T. Seaborg synthesized the sixth transuranium element to be made in a laboratory using the cyclotron at the University of California at Berkeley to bombard curium-242 with alpha particles (helium nuclei). They named the new element after the state of California.

The first isotope of californium to be made was californium-245, with a half-life of only 44 minutes. Some 14 other isotopes of the element have since been made, usually by subjecting berkelium to the intense neutron radiation produced in a nuclear reactor.

The most interesting isotope of californium to date has been californium-252, an isotope that spontaneously emits neutrons. With a half-life of 2.65 years, it is extremely radioactive and must be handled with great care.

Neutron sources are ordinarily hard to come by. Either a nuclear reactor is required to generate neutrons from an element, or some highly radioactive emitter of alpha particles (helium nuclei), such as plutonium, must be mixed with beryllium powder. The discovery of an extremely portable neutron source such as californium-252 suggests many possible applications for it. It can easily be taken into the field for the analysis of oil-bearing layers of earth or for the mining of gold and silver by activation. The activation process occurs because neutrons, having no electrical charge, are easily absorbed by the nuclei of other atoms, rather than being repelled by their electron shells.

The cyclotron at the University of California at Berkeley's Lawrence Berkeley Laboratory. A cyclotron uses a strong magnetic field and an electric field of alternating polarity to accelerate particles along a spiral path to high velocities.

Once their nuclei have been rendered radioactive in this way, it is often easier to identify gold or silver by examining their radiation characteristics than by analyzing them chemically.

The U.S. Nuclear Regulatory Commission, in an effort to encourage its use, is making californium-252 available in small quantities. Even seemingly microscopic quantities of the isotope are often sufficient for many applications, particularly the demonstrations of neutron reactions in college and university laboratories.

Einsteinium

Atomic Number 99

Chemical Symbol Es

Group IIIB—Transition Element (The Actinides)

IA	IA	IIIB	IVB	VB	VIB	VIIB	VIIIB			IB	IIB	IIIA	IVA	VA	VIA	VIIA	VIIIA
H																	He
Li	Be											B	C	N	O	F	Ne
Na	Mg											Al	Si	P	S	Cl	Ar
K	Ca	Sc	Ti	V	Cr	Mn	Fe	Co	Ni	Cu	Zn	Ga	Ge	As	Se	Br	Kr
Rb	Sr	Y	Zr	Nb	Mo	Tc	Ru	Rh	Pd	Ag	Cd	In	Sn	Sb	Te	I	Xe
Cs	Ba	*La	Hf	Ta	W	Re	Os	Ir	Pt	Au	Hg	Tl	Pb	Bi	Po	At	Rn
Fr	Ra	†Ac	Rf	Db	Sg	Bh	Hs	Mt	Ds	Rg	Cn	Uut	Uuq	Uup	Uuh	Uus	Uuo

*	Ce	Pr	Nd	Pm	Sm	Eu	Gd	Tb	Dy	Ho	Er	Tm	Yb	Lu
†	Th	Pa	U	Np	Pu	Am	Cm	Bk	Cf	Es	Fm	Md	No	Lr

Es

Einsteinium, named for the celebrated physicist Albert Einstein, was discovered by the American physicist Albert Ghiorso and his coworkers in 1952. They found the seventh transuranium element, subsequently called einsteinium-253, while investigating the debris of the first hydrogen bomb explosion in the Pacific Ocean. The sample they recovered weighed only one hundred millionth of a gram, and the isotope had a half-life of 20 days. The discovery was not announced until 1955 because information about atomic testing was restricted by the United States at that time.

Some 16 isotopes of einsteinium are now known. The most stable of these is einsteinium-252, with a half-life of 471 days. Most of the known isotopes of einsteinium have been produced in the High Flux Isotope Reactor at Oak Ridge National Laboratory in Tennessee by irradiating plutonium-239 with intense beams of neutrons. The plutonium samples are often left in the reactor for several months before the chemical separation of einsteinium can be efficiently performed. At Oak Ridge, the reactor used for producing transuranium elements lies beneath a large pool of water. It is possible to stand above it and see the eerie blue light produced by the high energy radiation from these elements.

Einsteinium is extremely radioactive, and great care must be taken when handling it.

Workers at Oak Ridge National Laboratory carefully remove plutonium samples from the High Flux Isotope Reactor.

Fermium

IA	IA	IIIB	IVB	VB	VIB	VIIB	VIIIB			IB	IIB	IIIA	IVA	VA	VIA	VIIA	VIIIA
H																	He
Li	Be											B	C	N	O	F	Ne
Na	Mg											Al	Si	P	S	Cl	Ar
K	Ca	Sc	Ti	V	Cr	Mn	Fe	Co	Ni	Cu	Zn	Ga	Ge	As	Se	Br	Kr
Rb	Sr	Y	Zr	Nb	Mo	Tc	Ru	Rh	Pd	Ag	Cd	In	Sn	Sb	Te	I	Xe
Cs	Ba	*La	Hf	Ta	W	Re	Os	Ir	Pt	Au	Hg	Tl	Pb	Bi	Po	At	Rn
Fr	Ra	†Ac	Rf	Db	Sg	Bh	Hs	Mt	Ds	Rg	Cn	Uut	Uuq	Uup	Uuh	Uus	Uuo

*	Ce	Pr	Nd	Pm	Sm	Eu	Gd	Tb	Dy	Ho	Er	Tm	Yb	Lu
†	Th	Pa	U	Np	Pu	Am	Cm	Bk	Cf	Es	Fm	Md	No	Lr

Fm

Like einsteinium, the element that precedes it in the periodic table, fermium, the eighth transuranium element, was identified in 1952 by the physicist Albert Ghiorso and his coworkers. As had also been the case with einsteinium, they discovered fermium-252, with a half-life of 20 hours, in the debris formed by the first hydrogen bomb explosion in the Pacific Ocean. They named the element in honor of the celebrated Italian-born physicist Enrico Fermi, who in 1944 had generated the first self-sustaining nuclear reaction in a facility located under the grandstand of the athletic field at the University of Chicago. For security reasons, the discovery of fermium was not announced until 1955.

Eighteen isotopes of fermium are now known to exist, all of which are intensely radioactive. They are usually synthesized by subjecting elements such as uranium and plutonium to intense neutron bombardment. In a neutron-rich environment such as the High Flux Isotope Reactor at Oak Ridge National Laboratory in Tennessee, or in the center of a thermonuclear explosion, an element such as uranium can undergo successive neutron capture, often absorbing as many as 16 or 17 neutrons, to produce the heavy transuranium elements.

The longest lived of the fermium isotopes is fermium-257, with a half-life of about 82 days. It can be produced in amounts large enough to be studied. No commercial or research applications for fermium have yet been developed.

Atomic Number **100**

Chemical Symbol **Fm**

Group **IIIB—Transition Element (The Actinides)**

IA																	VIIIA
H	IA											IIIA	IVA	VA	VIA	VIIA	He
Li	Be											B	C	N	O	F	Ne
Na	Mg	IIIB	IVB	VB	VIB	VIIB	VIIIB			IB	IIB	Al	Si	P	S	Cl	Ar
K	Ca	Sc	Ti	V	Cr	Mn	Fe	Co	Ni	Cu	Zn	Ga	Ge	As	Se	Br	Kr
Rb	Sr	Y	Zr	Nb	Mo	Tc	Ru	Rh	Pd	Ag	Cd	In	Sn	Sb	Te	I	Xe
Cs	Ba	*La	Hf	Ta	W	Re	Os	Ir	Pt	Au	Hg	Tl	Pb	Bi	Po	At	Rn
Fr	Ra	†Ac	Rf	Db	Sg	Bh	Hs	Mt	Ds	Rg	Cn	Uut	Uuq	Uup	Uuh	Uus	Uuo

*	Ce	Pr	Nd	Pm	Sm	Eu	Gd	Tb	Dy	Ho	Er	Tm	Yb	Lu
†	Th	Pa	U	Np	Pu	Am	Cm	Bk	Cf	Es	Fm	Md	No	Lr

Mendelevium

Atomic Number **101**

Chemical Symbol **Md**

Group IIIB—Transition Element (The Actinides)

Md

The ninth artificial transuranium element was discovered in 1955 by a team of scientists led by the physicist Albert Ghiorso at the University of California at Berkeley. Continuing their search for ever heavier elements, they used the cyclotron at Berkeley to bombard einsteinium-253 with alpha particles (helium nuclei) and eventually fabricated mendelevium-256, with a half-life of 77 minutes. The element was named for the Russian chemist Dmitry Mendeleyev, the creator of the periodic table. The small amounts of mendelevium-256 that were originally produced at Berkeley made its identification very difficult. It is often said that this element was synthesized one atom at a time.

Some 13 isotopes of mendelevium are now known. All are radioactive, with the longest lived being mendelevium-258, which has a half-life of 56 days. Only trace amounts of these isotopes have been made, and little is known of their chemistry. They have no known applications.

Nobelium

IA																	VIIIA
H	IA											IIIA	IVA	VA	VIA	VIIA	He
Li	Be											B	C	N	O	F	Ne
Na	Mg	IIIB	IVB	VB	VIB	VIIB	VIIIB			IB	IIB	Al	Si	P	S	Cl	Ar
K	Ca	Sc	Ti	V	Cr	Mn	Fe	Co	Ni	Cu	Zn	Ga	Ge	As	Se	Br	Kr
Rb	Sr	Y	Zr	Nb	Mo	Tc	Ru	Rh	Pd	Ag	Cd	In	Sn	Sb	Te	I	Xe
Cs	Ba	*La	Hf	Ta	W	Re	Os	Ir	Pt	Au	Hg	Tl	Pb	Bi	Po	At	Rn
Fr	Ra	†Ac	Rf	Db	Sg	Bh	Hs	Mt	Ds	Rg	Cn	Uut	Uuq	Uup	Uuh	Uus	Uuo

*	Ce	Pr	Nd	Pm	Sm	Eu	Gd	Tb	Dy	Ho	Er	Tm	Yb	Lu
†	Th	Pa	U	Np	Pu	Am	Cm	Bk	Cf	Es	Fm	Md	**No**	Lr

Atomic Number **102**

Chemical Symbol **No**

Group **IIIB—Transition Element (The Actinides)**

No

The sequence of events that led to the discovery of the transuranium element nobelium is somewhat confused. The work that led to the first claim of its discovery by a team of scientists working at the Nobel Institute in Stockholm was shown to be faulty. Subsequently, in 1958, the physicist Albert Ghiorso and his colleagues, working at the University of California at Berkeley, unambiguously identified nobelium-254, which has a half-life of 55 seconds.

In creating nobelium-254, the Berkeley group abandoned the cyclotron, the accelerator they had previously used so successfully to manufacture a host of transuranium elements. Instead, they bombarded a sample of curium-244 and curium-246 with carbon-12 ions using the Heavy Ion Linear Accelerator at Berkeley. Their success in synthesizing nobelium-254 was confirmed by a group of Russian physicists working at Dubna, in the Soviet Union. Ghiorso and his coworkers decided to retain the original name of the element, which had been assigned to it by the Stockholm group in honor of Alfred Nobel, the inventor of dynamite.

Eleven isotopes of nobelium have so far been synthesized, and all are radioactive. Nobelium-259 is the longest lived of the isotopes, with a half-life of 57 minutes. Nobelium has not been produced in quantities large enough to permit the study of its chemical and physical properties.

Lawrencium

IA	IA	IIIB	IVB	VB	VIB	VIIB	VIIIB			IB	IIB	IIIA	IVA	VA	VIA	VIIA	VIIIA
H																	He
Li	Be											B	C	N	O	F	Ne
Na	Mg											Al	Si	P	S	Cl	Ar
K	Ca	Sc	Ti	V	Cr	Mn	Fe	Co	Ni	Cu	Zn	Ga	Ge	As	Se	Br	Kr
Rb	Sr	Y	Zr	Nb	Mo	Tc	Ru	Rh	Pd	Ag	Cd	In	Sn	Sb	Te	I	Xe
Cs	Ba	*La	Hf	Ta	W	Re	Os	Ir	Pt	Au	Hg	Tl	Pb	Bi	Po	At	Rn
Fr	Ra	†Ac	Rf	Db	Sg	Bh	Hs	Mt	Ds	Rg	Cn	Uut	Uuq	Uup	Uuh	Uus	Uuo

*	Ce	Pr	Nd	Pm	Sm	Eu	Gd	Tb	Dy	Ho	Er	Tm	Yb	Lu
†	Th	Pa	U	Np	Pu	Am	Cm	Bk	Cf	Es	Fm	Md	No	Lr

Atomic Number 103

Chemical Symbol Lr

Group IIIB—Transition Element (The Actinides)

Lr

Continuing their astonishing string of discoveries at Berkeley, a team of scientists led by Albert Ghiorso synthesized and identified lawrencium, a new transuranium element heavier than nobelium, in 1961.

Using the Heavy Ion Linear Accelerator at the University of California, they bombarded a mixture of three isotopes of californium with boron-10 and boron-11 ions. The target weighed only a few millionths of a gram. Careful analysis of the reaction indicated that the Berkeley team had manufactured lawrencium-258, an isotope of lawrencium with a half-life of 4 seconds. It was named in honor of Ernest O. Lawrence, the inventor of the cyclotron and a former professor at the University of California.

Eight isotopes of lawrencium have been synthesized to date, with the longest lived being lawrencium-256, which has a half-life of about 30 seconds.

Very little is known about the chemical and physical properties of lawrencium, although Ghiorso and his colleagues, working with unbelievably small amounts of lawrencium equivalent to a few atoms, managed a very preliminary study of the oxidation behavior of the element. They found that the chemical behavior of lawrencium seemed to resemble that of the lighter actinides.

Element 103 was named lawrencium in honor of Ernest O. Lawrence, the inventor of the cyclotron.

Rutherfordium

Atomic Number **104**

Chemical Symbol **Rf**

Group **IVB—A Transactinide**

IA																	VIIIA
H	IA											IIIA	IVA	VA	VIA	VIIA	He
Li	Be											B	C	N	O	F	Ne
Na	Mg	IIIB	IVB	VB	VIB	VIIB		VIIIB		IB	IIB	Al	Si	P	S	Cl	Ar
K	Ca	Sc	Ti	V	Cr	Mn	Fe	Co	Ni	Cu	Zn	Ga	Ge	As	Se	Br	Kr
Rb	Sr	Y	Zr	Nb	Mo	Tc	Ru	Rh	Pd	Ag	Cd	In	Sn	Sb	Te	I	Xe
Cs	Ba	*La	Hf	Ta	W	Re	Os	Ir	Pt	Au	Hg	Tl	Pb	Bi	Po	At	Rn
Fr	Ra	†Ac	**Rf**	Db	Sg	Bh	Hs	Mt	Ds	Rg	Cn	Uut	Uuq	Uup	Uuh	Uus	Uuo

*	Ce	Pr	Nd	Pm	Sm	Eu	Gd	Tb	Dy	Ho	Er	Tm	Yb	Lu
†	Th	Pa	U	Np	Pu	Am	Cm	Bk	Cf	Es	Fm	Md	No	Lr

Rf

A history of competing claims confused the naming of element 104, the first of the elements beyond the actinides in the periodic table, which are usually called transactinide elements. The honor of naming a new element usually goes to the discoverer. When the discovery is disputed, a system of nomenclature designed by the International Union of Pure and Applied Chemistry (IUPAC) is used to identify the atomic number of the new element. For this purpose, IUPAC has recommended the use of the following code, which assigns a syllable to each digit in an element's atomic number, with all names ending in *ium*.

0 = nil	5 = pent
1 = un	6 = hex
2 = bi	7 = sept
3 = tri	8 = oct
4 = quad	9 = enn

The dispute over the discovery of element 104 began in 1964, when a team of Russian scientists working at the Joint Institute for Nuclear Research at Dubna, in the Soviet Union, reported that they had created a new element, unnilquadium-260, with a half-life of 0.3 of a second. The original, ungainly name of "un-nil-quad-ium" (1-0-4-ium) was based on the IUPAC system; its chemical symbol was Unq. In their experiments, the Russian scientists bombarded plutonium-242 with neon-22 ions. They subsequently suggested the name kurchatovium for the new element, in honor of the head of Soviet Research, Ivan Kurchatov. The evidence for this discovery, however, was not very convincing to the international community.

Soviet scientists claimed credit for discovering both rutherfordium (for which they proposed the name kurchatovium in honor of Soviet chemist Ivan Kurchatov, above) and hahnium, now named dubnium. However, their claims were rejected by the International Union of Pure and Applied Chemistry.

Then, in 1969, a team led by the American physicist Albert Ghiorso bombarded californium-249 with carbon-12 ions using the Heavy Ion Linear Accelerator at the University of California at Berkeley. They reported the positive identification of unnilquadium-257, an isotope with a half-life of 4 to 5 seconds. Large amounts of this isotope have subsequently been detected. The Berkeley group proposed naming the element rutherfordium, in honor of the distinguished New Zealand physicist Ernest Rutherford, whose work had been instrumental in the early understanding of the atom. The American claim won the day, and the name rutherfordium is now the name endorsed by the American Chemical Society.

Six isotopes of rutherfordium, all radioactive, have so far been identified. Rutherfordium-261, the longest lived, has a half-life of 62 seconds. Little is known about the chemical or physical properties of the element.

Dubnium

IA																	VIIIA
H	IA											IIIA	IVA	VA	VIA	VIIA	He
Li	Be											B	C	N	O	F	Ne
Na	Mg	IIIB	IVB	VB	VIB	VIIB		VIIIB		IB	IIB	Al	Si	P	S	Cl	Ar
K	Ca	Sc	Ti	V	Cr	Mn	Fe	Co	Ni	Cu	Zn	Ga	Ge	As	Se	Br	Kr
Rb	Sr	Y	Zr	Nb	Mo	Tc	Ru	Rh	Pd	Ag	Cd	In	Sn	Sb	Te	I	Xe
Cs	Ba	*La	Hf	Ta	W	Re	Os	Ir	Pt	Au	Hg	Tl	Pb	Bi	Po	At	Rn
Fr	Ra	†Ac	Rf	Db	Sg	Bh	Hs	Mt	Ds	Rg	Cn	Uut	Uuq	Uup	Uuh	Uus	Uuo

*	Ce	Pr	Nd	Pm	Sm	Eu	Gd	Tb	Dy	Ho	Er	Tm	Yb	Lu
†	Th	Pa	U	Np	Pu	Am	Cm	Bk	Cf	Es	Fm	Md	No	Lr

Disputed claims of its discovery have plagued element 105, previously known as unnilpentium (see the section on rutherfordium for a description of the naming process used for new elements). In 1967, a group of Russian scientists working at the Joint Institute for Nuclear Research at Dubna, in the Soviet Union, bombarded americium-243 with the heavy ions of neon-22. They claimed to have produced a few atoms each of unnilpentium-260 and unnilpentium-261.

In 1970, a team headed by the American physicist Albert Ghiorso bombarded californium-249 with heavy nitrogen-15 ions using the Heavy Ion Linear Accelerator at the University of California at Berkeley and positively identified unnilpentium-260, with a half-life of 1.6 seconds, among the products of the bombardment. Ghiorso and his colleagues attempted to duplicate the Russian experiment, but this proved fruitless. The Berkeley group proposed to name their new element hahnium in honor of Otto Hahn, the German chemist who discovered nuclear fission.

In view of the overwhelming evidence presented by the Berkeley team, the American Chemical Society officially endorsed the name of hahnium for unnilpentium. Ghiorso and his team continued their work with the element and produced two new isotopes in 1971. In 1997 the International Union of Pure and Applied Chemistry decided to change the name of this element to dubnium. There are now five known isotopes of dubnium, and all are radioactive. The longest lived is dubnium-262, with a half-life of 34 seconds. Its chemical and physical properties are unknown.

Atomic Number **105**

Chemical Symbol **Db**

Group **VB—A Transactinide**

Seaborgium

IA	IA	IIIB	IVB	VB	VIB	VIIB	VIIIB			IB	IIB	IIIA	IVA	VA	VIA	VIIA	VIIIA
H																	He
Li	Be											B	C	N	O	F	Ne
Na	Mg											Al	Si	P	S	Cl	Ar
K	Ca	Sc	Ti	V	Cr	Mn	Fe	Co	Ni	Cu	Zn	Ga	Ge	As	Se	Br	Kr
Rb	Sr	Y	Zr	Nb	Mo	Tc	Ru	Rh	Pd	Ag	Cd	In	Sn	Sb	Te	I	Xe
Cs	Ba	*La	Hf	Ta	W	Re	Os	Ir	Pt	Au	Hg	Tl	Pb	Bi	Po	At	Rn
Fr	Ra	†Ac	Rf	Db	Sg	Bh	Hs	Mt	Ds	Rg	Cn	Uut	Uuq	Uup	Uuh	Uus	Uuo

*	Ce	Pr	Nd	Pm	Sm	Eu	Gd	Tb	Dy	Ho	Er	Tm	Yb	Lu
†	Th	Pa	U	Np	Pu	Am	Cm	Bk	Cf	Es	Fm	Md	No	Lr

Atomic Number **106**

Chemical Symbol **Sg**

Group **VIB—A Transactinide**

Sg

Like the two preceding elements in the periodic table, the claim of discovery of element 106, along with the right to name it, was a subject of dispute. After a waiting period of some 10 years, an international group of referees finally approved the name unnilhexium for the new element (see the discussion of how new elements are named in the section on rutherfordium). In March 1994, the American Chemical Society announced that the new element was to be named seaborgium in honor of Glenn T. Seaborg, who was a member of the team that discovered this element and who won the 1951 Nobel Prize in chemistry for his work in the discovery of plutonium and nine other artificially created elements.

In June 1974, a team of Russian scientists working at the Joint Institute for Nuclear Research in Dubna, in what was then the Soviet Union, reported that they had produced unnilhexium by bombarding lead-206 with highly energetic heavy chromium-54 ions produced in their cyclotron. Because other experiments failed to confirm this result, their claim of discovery was in doubt.

At about the same time, a team of scientists at the Lawrence Livermore Laboratory and the University of California at Berkeley triumphantly reported the unambiguous discovery of unnilhexium-263, with a half-life of 0.9 of a second. In generating the element, the California scientists used the Berkeley Heavy Ion Linear Accelerator to bombard californium-249 with oxygen-18. They produced only a tiny amount of element 106, however, so that international regulatory groups delayed in resolving the conflicting claims of prior discovery of the new element.

The technique that was used by the California group to identify the element is quite fascinating. It employed an elaborate apparatus previously used to discover rutherfordium and hahnium, in which jets of air propelled the products of nuclear reactions to the top of a vertical wheel, where they were deposited. As the wheel rotated, detectors wired to computers and mounted around the wheel then identified these products by monitoring the half-lives of the decay products to which they gave rise.

In 1993, scientists at the Lawrence Livermore and Berkeley laboratories repeated their experiment, confirming the original result, and established their claim to the discovery of element 106. Four isotopes of seaborgium have been identified; seaborgium-263 is the one with the longest half-life.

It is interesting to note that the naming of seaborgium by a group at the University of California at Berkeley set off an international disagreement on the naming of many of the recently discovered elements. Although the honor of naming an element had always been granted to its discoverers, an international commission organized by the International Union of Pure and Applied Chemists (IUPAC) voted at the end of August 1994 to disallow the name on the grounds that Glenn T. Seaborg was still alive. The commission went even further and proposed an entirely new list of names for elements 104 to 108.

After 3 years of meetings and discussions, a compromise was finally agreed to by an international group of chemists during the summer of 1997. The names now accepted by the IUPAC are the ones used in this book.

In July 1997, a consortium of scientists led by Dr. Matthias Schadel at the Heavy Ion Research laboratory (GSI) in Darmstadt, Germany, reported that they had managed to do chemical analysis on seaborgium. This was the first time that any such work had ever been done, and it was done with an incredibly small sample of only seven seaborgium atoms. They found that seaborgium seemed to have chemical properties consistent with its position in the periodic table. That is, it produced reaction products similar to molybdenum and tungsten, the elements just above it in column 6 of the table. This came as a surprise, as elements 104 and 105 do not appear to react in ways consistent with their position in the table. It is as yet not known why seaborgium behaves differently from these two lighter atoms.

Bohrium

IA	IA	IIIB	IVB	VB	VIB	VIIB	VIIIB			IB	IIB	IIIA	IVA	VA	VIA	VIIA	VIIIA
H																	He
Li	Be											B	C	N	O	F	Ne
Na	Mg											Al	Si	P	S	Cl	Ar
K	Ca	Sc	Ti	V	Cr	Mn	Fe	Co	Ni	Cu	Zn	Ga	Ge	As	Se	Br	Kr
Rb	Sr	Y	Zr	Nb	Mo	Tc	Ru	Rh	Pd	Ag	Cd	In	Sn	Sb	Te	I	Xe
Cs	Ba	*La	Hf	Ta	W	Re	Os	Ir	Pt	Au	Hg	Tl	Pb	Bi	Po	At	Rn
Fr	Ra	†Ac	Rf	Db	Sg	Bh	Hs	Mt	Ds	Rg	Cn	Uut	Uuq	Uup	Uuh	Uus	Uuo

*	Ce	Pr	Nd	Pm	Sm	Eu	Gd	Tb	Dy	Ho	Er	Tm	Yb	Lu
†	Th	Pa	U	Np	Pu	Am	Cm	Bk	Cf	Es	Fm	Md	No	Lr

Bh

In 1976, a team of Russian scientists at the Joint Institute for Nuclear Research at Dubna in the Soviet Union announced that they had synthesized unnilseptium, element number 107. They claimed to have created the new element by bombarding bismuth-204 with heavy ions of chromium-54 that had been accelerated in a cyclotron. Their claim was rejected.

In 1981, physicists Peter Armbruster and Gottfried Munzenberg, working in Darmstadt, Germany, also announced the creation of unnilseptium. They proposed the name *nielsbohrium* for this element, in honor of the great Danish physicist Niels Bohr, whose work led to the modern concept of the atom. Their research claims were confirmed in 1992 by the International Union of Pure and Applied Chemistry, and the American Chemical Society approved the name the German team proposed for the new element. In 1997 the International Union of Pure and Applied Chemistry decided to change the name of this element to bohrium.

The chemical and physical properties of the isotopes of bohrium are unknown.

Atomic Number **107**

Chemical Symbol **Bh**

Group **VIIB—A Transactinide**

IA	IA	IIIB	IVB	VB	VIB	VIIB	VIIIB			IB	IIB	IIIA	IVA	VA	VIA	VIIA	VIIIA
H																	He
Li	Be											B	C	N	O	F	Ne
Na	Mg											Al	Si	P	S	Cl	Ar
K	Ca	Sc	Ti	V	Cr	Mn	Fe	Co	Ni	Cu	Zn	Ga	Ge	As	Se	Br	Kr
Rb	Sr	Y	Zr	Nb	Mo	Tc	Ru	Rh	Pd	Ag	Cd	In	Sn	Sb	Te	I	Xe
Cs	Ba	*La	Hf	Ta	W	Re	Os	Ir	Pt	Au	Hg	Tl	Pb	Bi	Po	At	Rn
Fr	Ra	†Ac	Rf	Db	Sg	Bh	Hs	Mt	Ds	Rg	Cn	Uut	Uuq	Uup	Uuh	Uus	Uuo

*	Ce	Pr	Nd	Pm	Sm	Eu	Gd	Tb	Dy	Ho	Er	Tm	Yb	Lu
†	Th	Pa	U	Np	Pu	Am	Cm	Bk	Cf	Es	Fm	Md	No	Lr

Hassium

In 1984, a team led by physicists Peter Armbruster and Gottfried Munzenberg, working in Darmstadt, Germany, announced the discovery of unniloctium, element 108. This was the same team that had synthesized bohrium.

The name that the discoverers proposed for this new element was hassium, for *hassia*, the Latin name for the German state of Hesse. In 1992, the International Union of Pure and Applied Chemistry confirmed the research claims of the German team and the American Chemical Society endorsed the proposed name. The chemical and physical properties of hassium are unknown.

The German physicists Peter Armbruster (back row, second from right) and Gottfried Munzenberg (front) led the team that discovered elements 107, 108, 109, 110, and 111.

Atomic Number **108**

Chemical Symbol **Hs**

Group **VIIIB—A Transactinide**

Meitnerium

IA																	VIIIA
H	IA											IIIA	IVA	VA	VIA	VIIA	He
Li	Be											B	C	N	O	F	Ne
Na	Mg	IIIB	IVB	VB	VIB	VIIB	VIIIB			IB	IIB	Al	Si	P	S	Cl	Ar
K	Ca	Sc	Ti	V	Cr	Mn	Fe	Co	Ni	Cu	Zn	Ga	Ge	As	Se	Br	Kr
Rb	Sr	Y	Zr	Nb	Mo	Tc	Ru	Rh	Pd	Ag	Cd	In	Sn	Sb	Te	I	Xe
Cs	Ba	*La	Hf	Ta	W	Re	Os	Ir	Pt	Au	Hg	Tl	Pb	Bi	Po	At	Rn
Fr	Ra	†Ac	Rf	Db	Sg	Bh	Hs	Mt	Ds	Rg	Cn	Uut	Uuq	Uup	Uuh	Uus	Uuo

*	Ce	Pr	Nd	Pm	Sm	Eu	Gd	Tb	Dy	Ho	Er	Tm	Yb	Lu
†	Th	Pa	U	Np	Pu	Am	Cm	Bk	Cf	Es	Fm	Md	No	Lr

Mt

In 1982, the team of physicists that had discovered transactinide elements 107 and 108, led by Peter Armbruster and Gottfried Munzenberg, working in Darmstadt, Germany, announced the synthesis and discovery of element 109, unnilennium. In the research that led to the discovery of this new element, they bombarded a bismuth-209 target with high-energy iron-58 ions to create unnilennium-266. Incredible as it might seem, only three atoms of unnilennium were created, and they decayed after only 3.4 thousandths of a second. This was a very brief existence but long enough to identify their structure.

The German team confirmed the existence of unnilennium by following the series of decay products to which it gave rise. For the new element, they proposed the name *meitnerium*, in honor of Lise Meitner, the German physicist who, along with her nephew, Otto R. Frisch, had first envisioned nuclear fission as a splitting of the uranium nucleus.

In 1992, the International Union of Pure and Applied Chemistry confirmed the German team's research claims and both the IUPAC and the American Chemical Society approved the proposed name. It's chemical and physical properties remain unknown.

Atomic Number **109**

Chemical Symbol **Mt**

Group **VIIIB—A Transactinide**

IA																	VIIIA
H	IA											IIIA	IVA	VA	VIA	VIIA	**He**
Li	**Be**											**B**	**C**	**N**	**O**	**F**	**Ne**
Na	**Mg**	IIIB	IVB	VB	VIB	VIIB		VIIIB		IB	IIB	**Al**	**Si**	**P**	**S**	**Cl**	**Ar**
K	**Ca**	**Sc**	**Ti**	**V**	**Cr**	**Mn**	**Fe**	**Co**	**Ni**	**Cu**	**Zn**	**Ga**	**Ge**	**As**	**Se**	**Br**	**Kr**
Rb	**Sr**	**Y**	**Zr**	**Nb**	**Mo**	**Tc**	**Ru**	**Rh**	**Pd**	**Ag**	**Cd**	**In**	**Sn**	**Sb**	**Te**	**I**	**Xe**
Cs	**Ba**	***La**	**Hf**	**Ta**	**W**	**Re**	**Os**	**Ir**	**Pt**	**Au**	**Hg**	**Tl**	**Pb**	**Bi**	**Po**	**At**	**Rn**
Fr	**Ra**	**†Ac**	**Rf**	**Db**	**Sg**	**Bh**	**Hs**	**Mt**	**Ds**	**Rg**	**Cn**	**Uut**	**Uuq**	**Uup**	**Uuh**	**Uus**	**Uuo**

*	**Ce**	**Pr**	**Nd**	**Pm**	**Sm**	**Eu**	**Gd**	**Tb**	**Dy**	**Ho**	**Er**	**Tm**	**Yb**	**Lu**
†	**Th**	**Pa**	**U**	**Np**	**Pu**	**Am**	**Cm**	**Bk**	**Cf**	**Es**	**Fm**	**Md**	**No**	**Lr**

Darmstadtium

Atomic Number **110**

Chemical Symbol **Ds**

Group **VIIIB—A Transactinide**

Ds

On November 14, 1994, an international team of heavy ion physicists, led by Peter Armbruster and Sigurd Hofmann, announced their discovery of element 110. After a search that took almost 10 years, the team created the new element by bombarding a target of lead-208 with a beam of nickel-62 ions that had been accelerated to a fraction of the speed of light. The lead was mounted on a rotating wheel to allow it to cool and prevent the lead from melting. A small number of the collisions led to the fusion of the nuclei to form the nucleus of element 110 with a mass number of 269.

The work was done in Darmstadt, Germany, at the GSI Helmholtz Center for Heavy Ion Research. The GSI facility, with its unique accelerator facility for heavy ion beams, has become one of the world's leading centers for the creation of new elements. It is also credited with the creation of elements 107, 108, 109, 111, and 112.

In August 2003, the Council of the International Union of Pure and Applied Chemistry officially approved the name darmstadtium, with the symbol Ds, for element 110. The name was chosen in honor of the German city where the element was first discovered.

Since its original discovery, some 15 isotopes of darmstadtium have been made in laboratories throughout the world. These isotopes are all extremely unstable and quickly break apart and decay into lighter elements, often within thousandths of a second. The isotope discovered by Armbruster and his group, darmstadtium-269, has a half-life of 270 millionths of a second. The most stable of the known isotopes is darmstadtium-281 with a half-life of 1.1 minutes.

Members of the group that discovered six new isotopes of the superheavy elements pose in front of the Lawrence Berkeley National Laboratory's 88-inch Cyclotron.

Roentgenium

IA	IA	IIIB	IVB	VB	VIB	VIIB	VIIIB			IB	IIB	IIIA	IVA	VA	VIA	VIIA	VIIIA
H																	He
Li	Be											B	C	N	O	F	Ne
Na	Mg											Al	Si	P	S	Cl	Ar
K	Ca	Sc	Ti	V	Cr	Mn	Fe	Co	Ni	Cu	Zn	Ga	Ge	As	Se	Br	Kr
Rb	Sr	Y	Zr	Nb	Mo	Tc	Ru	Rh	Pd	Ag	Cd	In	Sn	Sb	Te	I	Xe
Cs	Ba	*La	Hf	Ta	W	Re	Os	Ir	Pt	Au	Hg	Tl	Pb	Bi	Po	At	Rn
Fr	Ra	†Ac	Rf	Db	Sg	Bh	Hs	Mt	Ds	Rg	Cn	Uut	Uuq	Uup	Uuh	Uus	Uuo

*	Ce	Pr	Nd	Pm	Sm	Eu	Gd	Tb	Dy	Ho	Er	Tm	Yb	Lu
†	Th	Pa	U	Np	Pu	Am	Cm	Bk	Cf	Es	Fm	Md	No	Lr

Atomic Number **111**

Chemical Symbol **Rg**

Group **IB—A Transactinide**

Rg

On December 8, 1994, approximately 1 month after announcing the creation of element 110, a team of scientists led by Sigurd Hofmann working at the GSI Helmholtz Center for Heavy Ion Research in Darmstadt, Germany, created element 111. It had 111 protons and 161 neutrons in its nucleus, giving it a mass number of 272.

The International Union of Pure and Applied Chemistry officially approved the name roentgenium for the new element, with the symbol Rg, on November 1, 2004. It was named for Wilhelm Conrad Roentgen, the discoverer of X rays, following a long tradition of naming elements to honor famous scientists.

Roentgenium was created by bombarding a bismuth-209 target with a beam of fast-moving nickel-64 atoms that had been accelerated in GSI's huge linear accelerator. The bismuth had been evaporated onto a rotating wheel to prevent overheating. After irradiating the target for approximately 6 days, the team finally succeeded in fusing the nuclei of a small number of these atoms to produce three massive roentgenium-272 atoms. Like all superheavy atoms, it is very unstable, decaying into smaller atoms by emitting alpha particles after roughly 1.5 milliseconds (0.0015 seconds). The three decay chains observed unambiguously identified the parent atom as roentgenium. Several other isotopes have been created since the initial discovery, with roentgenium-281 being the most stable with a lifetime of approximately 23 seconds.

The quantities of roentgenium that have been produced have been too small to investigate its chemical properties, but since it lies in the same column as gold and silver, it is presumably a metal as well.

IA	IA	IIIB	IVB	VB	VIB	VIIB	VIIIB			IB	IIB	IIIA	IVA	VA	VIA	VIIA	VIIIA
H																	He
Li	Be											B	C	N	O	F	Ne
Na	Mg											Al	Si	P	S	Cl	Ar
K	Ca	Sc	Ti	V	Cr	Mn	Fe	Co	Ni	Cu	Zn	Ga	Ge	As	Se	Br	Kr
Rb	Sr	Y	Zr	Nb	Mo	Tc	Ru	Rh	Pd	Ag	Cd	In	Sn	Sb	Te	I	Xe
Cs	Ba	*La	Hf	Ta	W	Re	Os	Ir	Pt	Au	Hg	Tl	Pb	Bi	Po	At	Rn
Fr	Ra	†Ac	Rf	Db	Sg	Bh	Hs	Mt	Ds	Rg	Cn	Uut	Uuq	Uup	Uuh	Uus	Uuo

*	Ce	Pr	Nd	Pm	Sm	Eu	Gd	Tb	Dy	Ho	Er	Tm	Yb	Lu
†	Th	Pa	U	Np	Pu	Am	Cm	Bk	Cf	Es	Fm	Md	No	Lr

Copernicium

On February 21, 1996, only a little more than a year after creating element 111, a team of more than 20 international scientists led by Sigurd Hofmann at the GSI Helmholz Center for Heavy Ion Research announced their creation of element 112.

After an unusually long waiting period, the International Union of Pure and Applied Chemistry approved the name copernicium for element 112 on February 19, 2010. Its chemical symbol was designated as Cn. The name is in honor of the Polish astronomer Nicolaus Copernicus.

The research team created copernicium by bombarding a target of lead-208 atoms with a beam of zinc-70 ions. The zinc atoms were extremely energetic, having been accelerated to extremely high speeds by GSI's 120-meter-long linear accelerator. To dissipate the heat produced by the collisions, the target was prepared by depositing the lead on a rotating circular wheel. Knowing that the probability of two nuclei fusing after a collision was small, the target was irradiated for several weeks with a beam intensity of trillions of atoms per second.

One atom of copernicium was finally observed. The nucleus of the newly created element 112 had a mass number of 277. It was formed by the fusion of zinc with 30 protons and lead with 82 protons. Like all superheavy elements, it was extremely unstable, decaying within 280 millionths of a second into two decay chains by emission of alpha particles. A study of these decay chains was critical in identifying copernicium.

Several isotopes of copernicium have been identified by other laboratories, all of them unstable. The most stable isotope discovered to date, copernicium-285, has a half-life of 29 seconds.

The quantities of copernicium that have been produced to date are too small to determine any of its chemical properties.

Atomic Number 112

Chemical Symbol Cn

Group IIB—A Transactinide

Ununtrium

IA	IA	IIIB	IVB	VB	VIB	VIIB	VIIIB			IB	IIB	IIIA	IVA	VA	VIA	VIIA	VIIIA
H																	He
Li	Be											B	C	N	O	F	Ne
Na	Mg											Al	Si	P	S	Cl	Ar
K	Ca	Sc	Ti	V	Cr	Mn	Fe	Co	Ni	Cu	Zn	Ga	Ge	As	Se	Br	Kr
Rb	Sr	Y	Zr	Nb	Mo	Tc	Ru	Rh	Pd	Ag	Cd	In	Sn	Sb	Te	I	Xe
Cs	Ba	*La	Hf	Ta	W	Re	Os	Ir	Pt	Au	Hg	Tl	Pb	Bi	Po	At	Rn
Fr	Ra	†Ac	Rf	Db	Sg	Bh	Hs	Mt	Ds	Rg	Cn	Uut	Uuq	Uup	Uuh	Uus	Uuo

*	Ce	Pr	Nd	Pm	Sm	Eu	Gd	Tb	Dy	Ho	Er	Tm	Yb	Lu
†	Th	Pa	U	Np	Pu	Am	Cm	Bk	Cf	Es	Fm	Md	No	Lr

Atomic Number **113**

Chemical Symbol **Uut**

Group **IIIA—A Transactinide**

A collaboration between Russian scientists and American scientists from the Lawrence Livermore National Laboratory in California, led by Yuri Oganessian, announced on February 1, 2004, that they had identified a new superheavy element, element 113. They did their work at the Joint Institute for Nuclear Research in Dubna, Russia. A few months later, on September 28, 2004, a group of Japanese scientists working at the Cyclotron Center of RIKEN, the Institute of Physical and Chemical Research in Japan, supported the Dubna claim by announcing that they had also observed an atom of element 113.

The Dubna team created element 113 by bombarding atoms of americium with fast-moving atoms of calcium that had been accelerated to high speeds in their cyclotron. They managed to fuse the nuclei of these calcium and americium atoms to first form element 115, which then quickly decayed to form element 113. The Japanese team at RIKEN created element 113 using their cyclotron to fuse atoms of bismuth and zinc.

The new element has not been officially named as yet, but it is known as ununtrium (1-1-3-ium). Its symbol is Uut. There is some controversy about an official internationally approved name, which traditionally is chosen by the group that discovers the element. The Dubna team has suggested becquerelium in honor of Henri Becquerel, the French physicist, whereas the Japanese group has proposed either japonium or rikenium.

A number of isotopes of ununtrium have been synthesized since 2004 and, like all the superheavy elements, are unstable. The most stable isotope found to date is ununtrium-286, with a half-life of 28.3 seconds.

Relatively little is known about ununtrium since it has been produced in such small quantities.

Ununquadium

IA	IA	IIIB	IVB	VB	VIB	VIIB	VIIIB			IB	IIB	IIIA	IVA	VA	VIA	VIIA	VIIIA
H																	He
Li	Be											B	C	N	O	F	Ne
Na	Mg											Al	Si	P	S	Cl	Ar
K	Ca	Sc	Ti	V	Cr	Mn	Fe	Co	Ni	Cu	Zn	Ga	Ge	As	Se	Br	Kr
Rb	Sr	Y	Zr	Nb	Mo	Tc	Ru	Rh	Pd	Ag	Cd	In	Sn	Sb	Te	I	Xe
Cs	Ba	*La	Hf	Ta	W	Re	Os	Ir	Pt	Au	Hg	Tl	Pb	Bi	Po	At	Rn
Fr	Ra	†Ac	Rf	Db	Sg	Bh	Hs	Mt	Ds	Rg	Cn	Uut	**Uuq**	Uup	Uuh	Uus	Uuo

*	Ce	Pr	Nd	Pm	Sm	Eu	Gd	Tb	Dy	Ho	Er	Tm	Yb	Lu
†	Th	Pa	U	Np	Pu	Am	Cm	Bk	Cf	Es	Fm	Md	No	Lr

A team of scientists working at Dubna, the Joint Institute for Nuclear Research near Moscow in Russia, announced in January 1999 that they had created a new ultraheavy atom, element 114. Under the leadership of Yuri Oganessian, a Russian nuclear physicist, the team of 18 Russian and 5 American scientists identified a single atom of the long-sought-for new element. Experiments performed in 2009 by scientists at the Lawrence Berkeley National Laboratory in California confirmed the existence of the new superheavy element.

The experiment that produced the new element used the large Russian cyclotron housed at Dubna to bombard a rare neutron-enriched form of plutonium, plutonium-244, with a beam of calcium-48 nuclei. The American team supplied both of these rare radioactive isotopes. After some 40 days of bombardment, a calcium nucleus with its 20 protons fused with a plutonium nucleus with 94 protons, producing an atom with 114 protons and 184 neutrons. The reaction is generally termed a "hot" fusion reaction because of the high energies involved.

The creation of element 114, tentatively named ununquadium, is an important step forward in the study of extremely heavy elements. Although unstable, as are all the superheavy elements, ununquadium survived a relatively long time for such a heavy element. When it decayed it formed a chain of three successive unstable daughter isotopes, elements 112, 110, and 108. The measurements of the time intervals between these three successive decays followed by spontaneous fission was estimated to be the relatively long half-lives of about 30 seconds, then 10, 1, and 11 minutes, respectively, for element 114 and its daughters.

Atomic Number **114**

Chemical Symbol **Uuq**

Group **IVA—A Transactinide**

Ununhexium

IA																	VIIIA
H	IA											IIIA	IVA	VA	VIA	VIIA	He
Li	Be											B	C	N	O	F	Ne
Na	Mg	IIIB	IVB	VB	VIB	VIIB	VIIIB			IB	IIB	Al	Si	P	S	Cl	Ar
K	Ca	Sc	Ti	V	Cr	Mn	Fe	Co	Ni	Cu	Zn	Ga	Ge	As	Se	Br	Kr
Rb	Sr	Y	Zr	Nb	Mo	Tc	Ru	Rh	Pd	Ag	Cd	In	Sn	Sb	Te	I	Xe
Cs	Ba	*La	Hf	Ta	W	Re	Os	Ir	Pt	Au	Hg	Tl	Pb	Bi	Po	At	Rn
Fr	Ra	†Ac	Rf	Db	Sg	Bh	Hs	Mt	Ds	Rg	Cn	Uut	Uuq	Uup	Uuh	Uus	Uuo

*	Ce	Pr	Nd	Pm	Sm	Eu	Gd	Tb	Dy	Ho	Er	Tm	Yb	Lu
†	Th	Pa	U	Np	Pu	Am	Cm	Bk	Cf	Es	Fm	Md	No	Lr

In December 2000, a collaboration of scientists from the Lawrence Livermore National Laboratory in California and the Joint Institute for Nuclear Research in Dubna, Russia, announced that they had identified a single atom of element 116. The work, done in Dubna, was under the leadership of Yuri Oganessian and Ken Moody.

The new element has not been officially named, but it is known as ununhexium (1-1-6-ium), according to the system designated by the International Union of Pure and Applied Chemistry for naming elements (see rutherfordium). Its chemical symbol has been designated as Uuh.

The Dubna team created ununhexium by accelerating calcium atoms in their cyclotron to high speeds and energies and using them to bombard atoms of curium. They managed to fuse the nuclei of these two atoms to create ununhexium-292. The new element was highly unstable and decayed within about 49 milliseconds (0.049 seconds).

The existence of ununhexium has since been confirmed by other experiments performed at Dubna in 2004 and 2006. On June 1, 2011 The International Union of Pure and Applied Chemistry formally recognized the existence of ununhexium and established the Dubna group's priority for its discovery. The team now has the privilege of naming this new element.

In 1999, almost simultaneously with the Dubna announcement, a team of American scientists at the Lawrence Livermore National Laboratory in Berkeley, California, also claimed recognition for the discovery of ununhexium. They announced that they had discovered two new elements, element 118 and element

Atomic Number 116

Chemical Symbol Uuh

Group VIA—A
Transactinide

116. Subsequent experiments failed to confirm these major discoveries, and in July 2001, the crestfallen team wrote a note in *Physics Review Letters* retracting their announcement.

There are some four isotopes of ununhexium known to date and all are extremely unstable, with the most stable being ununhexium-293 with a half-life of 61 milliseconds (0.061 seconds). The amount of ununhexium produced to date is far too small to determine any of its chemical properties or possible uses. Its position in Group VIA, however, under tellurium and polonium, predicts that it should share some of their properties.

Ununseptium

IA	IA	IIIB	IVB	VB	VIB	VIIB	VIIIB			IB	IIB	IIIA	IVA	VA	VIA	VIIA	VIIIA
H																	He
Li	Be											B	C	N	O	F	Ne
Na	Mg											Al	Si	P	S	Cl	Ar
K	Ca	Sc	Ti	V	Cr	Mn	Fe	Co	Ni	Cu	Zn	Ga	Ge	As	Se	Br	Kr
Rb	Sr	Y	Zr	Nb	Mo	Tc	Ru	Rh	Pd	Ag	Cd	In	Sn	Sb	Te	I	Xe
Cs	Ba	*La	Hf	Ta	W	Re	Os	Ir	Pt	Au	Hg	Tl	Pb	Bi	Po	At	Rn
Fr	Ra	†Ac	Rf	Db	Sg	Bh	Hs	Mt	Ds	Rg	Cn	Uut	Uuq	Uup	Uuh	**Uus**	Uuo

*	Ce	Pr	Nd	Pm	Sm	Eu	Gd	Tb	Dy	Ho	Er	Tm	Yb	Lu
†	Th	Pa	U	Np	Pu	Am	Cm	Bk	Cf	Es	Fm	Md	No	Lr

Atomic Number 117

Chemical Symbol Uus

Group VIIA—A Transactinide

Uus

Eight years after the discovery of ununoctium, an international team of scientists from the Joint Institute for Nuclear Research in Dubna, Russia, the Lawrence Livermore National Laboratory in California, the Oak Ridge National Laboratory, in Tennessee, Vanderbilt University in Tennessee, the University of Nevada, Las Vegas, and the Research Institute of Atomic Reactors in Dimitrovgrad, Russia, announced the discovery of element 117 on April 9, 2010. It was a monumental achievement that Yuri Ognessian, the team leader, called "the culmination of a decade-long journey to expand the periodic table and write the next chapter in heavy element research."

The experiment that finally led to the discovery of element 117 was a difficult one that took several years to complete and proved to be an example of the fruitful cooperation of formerly competitive laboratories on two continents.

The method that was used to create element 117, the only missing element in the last row of the periodic table, was to bombard berkelium-249 with projectiles of neutron-rich calcium-48 isotopes that had been accelerated to a fraction of the speed of light.

The scarcity of berkelium was one of the reasons that the experiment was so problematic. It required some 250 days in the High Flux Isotope Reactor at the Oak Ridge National Laboratory to fabricate the 22.2 mg of the berkelium-249 used as the target. To compound the difficulty, this rare isotope is radioactive with a half-life of 314 days, which introduced some time restraints on the experiment.

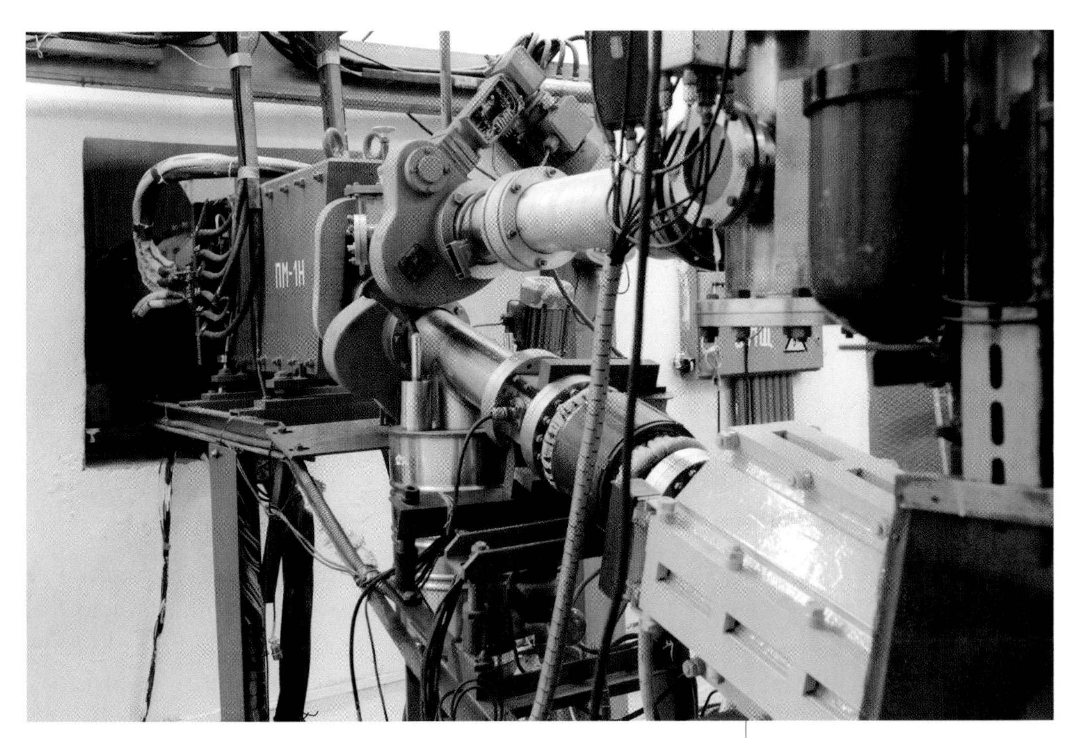

Equipment that is part of the particle accelerator at the Joint Institute for Nuclear Research in Dubna, Russia. Nuclei and heavy ions are accelerated to near the speed of light and made to collide with a stationary target.

After being purified and deposited onto arc-shape targets, the berkelium was placed inside the heavy ion U400 cyclotron at the Dubna facility and bombarded with calcium-48 isotopes for approximately 150 days. Here, fusion reactions resulting from a collision of the two nuclei produced two isotopes of element 117 with mass numbers 293 and 294. Both of these isotopes are unstable with half-lives of 21 ms (0.021 seconds) and 45 ms (0.045 seconds), respectively.

The two newly created isotopes are neutron rich, containing 173 and 174 neutrons, respectively, and come close to a magic number of 184 neutrons that the nuclear shell model predicts should form an island of stability. And indeed, the decay products of the two new isotopes are observed to have enhanced lifetimes and strongly suggest experimental support to this theory.

The new element has not been officially named, but is tentatively known as ununseptium (1-1-7-ium), according to the system designated by the International Union of Applied Chemistry for naming new elements (see the section on rutherfordium). It has been assigned a chemical symbol of Uus.

The quantities of ununseptium that have been produced to date are too small to measure any of its chemical properties or to predict any possible uses. But it has been of great theoretical use in furthering the believe that the predicted "island of stability" does, in fact, exist.

Ununoctium

IA	IA	IIIB	IVB	VB	VIB	VIIB	VIIIB			IB	IIB	IIIA	IVA	VA	VIA	VIIA	VIIIA
H																	He
Li	Be											B	C	N	O	F	Ne
Na	Mg											Al	Si	P	S	Cl	Ar
K	Ca	Sc	Ti	V	Cr	Mn	Fe	Co	Ni	Cu	Zn	Ga	Ge	As	Se	Br	Kr
Rb	Sr	Y	Zr	Nb	Mo	Tc	Ru	Rh	Pd	Ag	Cd	In	Sn	Sb	Te	I	Xe
Cs	Ba	*La	Hf	Ta	W	Re	Os	Ir	Pt	Au	Hg	Tl	Pb	Bi	Po	At	Rn
Fr	Ra	†Ac	Rf	Db	Sg	Bh	Hs	Mt	Ds	Rg	Cn	Uut	Uuq	Uup	Uuh	Uus	Uuo

*	Ce	Pr	Nd	Pm	Sm	Eu	Gd	Tb	Dy	Ho	Er	Tm	Yb	Lu
†	Th	Pa	U	Np	Pu	Am	Cm	Bk	Cf	Es	Fm	Md	No	Lr

In 1999, a group at the Lawrence Berkeley National Laboratory made headline news when they announced that they had discovered element 118 as well as element 116, also a previously unknown element. The excitement soon abated. These discoveries proved to be illusory, and after an investigation revealed that one of the investigators had fabricated data, the laboratory published a retraction of the discovery in 2001.

The unfortunate and embarrassing role played by element 118 in the search for new superheavy elements was to some extent mitigated by the announcement in 2002 by an international collaboration, led by Yuri Oganessian, of some 30 Russian and American scientists from the Lawrence Livermore National Laboratory in California that they had finally synthesized the heaviest known atom, element 118. Further experiments validating the existence of element 118 were again made between February and June 2005 and the results were published in the prestigious *Physical Review C* in October 2006.

The experiments were performed at the Joint Institute for Nuclear Research in Dubna, Russia, and done with great care to remove any possible doubts after the incident at Berkeley. A beam consisting of billions upon billions of calcium-48 ions was accelerated almost to the speed of light in Dubna's U400 heavy ion accelerator and made to strike a californium-249 target. The probability that these two atoms would collide and fuse was very low, so the target was irradiated for more than 96 days. Finally the group was able to detect a total of three atoms that signaled the existence of element 118.

Atomic Number **118**

Chemical Symbol **Uuo**

Group **VIIIA—A Transactinide**

An experiment storage ring at the GSI Helmholtz Center for Heavy Ion Research used to store particles that have been accelerated to high speeds in a cyclotron for use in experiments to search for superheavy elements.

Dubna's gas-filled recoil separator played a major role in weeding out and separating the 118 atoms from all the other particles and atoms in the target chamber. The newly created element has a mass number of 294 and is extremely unstable, decaying within thousands of a second down a chain to such elements as ununhexium and ununquadium. The transitions to these isotopes can be identified by their unique decay patterns and were critical in identifying element 118.

The new element has not been officially named as yet, a process that often takes years. Its tentative name is ununoctium (1-1-8-ium), according to the system designated by the International Union of Pure and Applied Chemistry for naming elements (see the section on rutherfordium). Its chemical symbol has been designated as Uuo.

The amount of ununoctium that has been produced to date is much too small to measure any of its chemical properties. Its position in the periodic table, however, is directly below radon, a radioactive noble gas, known, as are all the elements in this column, for being unreactive and inert. One would therefore expect ununoctium to share some of these properties.

Although it is far too early to predict any practical use for the element, scientists feel that great theoretical progress has been made in approaching the "island of stability" and a more profound understanding of the nature of matter.

Epilogue

Elements for a New Century

The announcement in 2006 that element 118, ununoctium, had finally been created was a major milestone reached in the world of superheavy atom physics. Not only did an international group working at the Joint Institute for Nuclear Research in Dubna, Russia, create the largest known atom, but also it helped to dissipate, to some extent, the embarrassment of the Berkeley Livermore group that had erroneously claimed the discovery of the element in 1999. The search for even heavier elements does not end here, however, and despite increasing technical difficulties, the work continues with renewed energy.

"Is there any limit to the size of a nucleus?" and "Is there really an island of stability?" are questions that major scientific establishments all over the world are trying to answer.

The island of stability is the name given to the region of the periodic table where the giant atoms being created will be unusually stable. Its existence is predicted on the basis of theories that say that when a nucleus contains 184 neutrons and has either 120 or 126 protons, so-called "magic numbers," it will have enhanced stability. The challenge is to create such a doubly magic nucleus, with magic numbers of protons and neutrons.

The motivation to continue this major effort to produce monster nuclei is that the answers being sought are of fundamental importance to our understanding of the universe. Demonstrating the existence of an island of stability, which has been discussed and theorized about for more than 40 years, would, in the opinion of many scientists, validate the theoretical work of a whole generation of physicists.

There is little doubt in the mind of scientists that this island really exists and that reaching it is an attainable goal. Demonstrating this will require experimenters of great imagination and almost virtuoso-like skill. Laboratories such as the Lawrence Livermore National Laboratory in Berkeley, California, the GSI Helmholtz Center for Heavy Ion Research in Darmstadt, Germany, and the Joint Institute for Nuclear Research in Dubna, Russia, are actively competing in the search by developing new techniques to fuse lighter elements that contain extremely large numbers of neutrons to counterbalance the repulsive force of the magic numbers of protons. New accelerating machines and detectors are constantly being developed, often at great expense. The driving force behind this effort to create superheavy atoms remains the search for knowledge that will initiate a rich new field of study of the nuclear and chemical properties of the elements.

There is also a more utilitarian component that motivates the search for the elements that make up this stable island. Many scientists believe, for example, that these new elements will form unusual materials with exotic properties never before seen. There is the hope that they might possibly have many practical uses in industry and technology and perhaps even serve as a source of energy in the form of new nuclear fuels.

The motivation to continue the major effort to produce monster nuclei is that the answers being sought are of fundamental importance to our understanding of the universe.

Glossary

Acid Any substance that produces hydrogen ions when dissolved in water.

Acid rain Rain that is made acidic when such pollutants as sulfur dioxide and nitrogen oxides are present in the atmosphere.

Adsorption The attraction of one substance to the surface of another.

Allotrope One of several possible physically distinct forms of an element.

Alloy A metallic substance that is either a compound or a mixture.

Alpha particle A particle consisting of two neutrons and two protons that is emitted by certain radioactive substances. It is essentially a helium nucleus.

Anion A negatively charged ion.

Atom The smallest and most basic unit of an element.

Atomic number The number of protons in the nucleus of one atom of an element.

Atomic weight The average weight of all the isotopes of an element.

Base A substance that produces hydroxyl ions $(OH)^-$ when dissolved in water.

Beta particle The electron emitted during the radioactive decay of certain radioisotopes.

Catalyst A substance that increases the rate of a chemical reaction without being changed chemically itself.

Cation A positively charged ion.

Compound A substance composed of two or more elements chemically bound together in a fixed ratio.

Corrosion The oxidation of metals in the atmosphere.

Crystal An important structure of certain solids in which the atoms or molecules that are its basic building blocks are arranged in regular repeating intervals.

Cyclotron A machine that accelerates nuclear particles to extremely high speeds. It is used to investigate the nature of matter and to form new elements.

Deuteron The nucleus of the isotope of hydrogen known as deuterium. It contains one proton and one neutron.

Electrolysis A process in which the passage of electric current through a cell causes a chemical reaction to occur.

Element A substance that cannot be decomposed into a simpler substance by any chemical or physical reaction.

Gamma rays A form of electromagnetic energy given off by certain radioactive atoms. They resemble X rays in their great penetrating power.

Group The elements that make up a column in the periodic table.

Half-life For radioactive elements, the time required for half of the element to decay.

Ion An atom or group of chemically bound atoms that has either a positive or a negative electrical charge.

Island of stability A theory that postulates the existence of superheavy elements with unusually long half-lives.

Isotopes Atoms of the same element that contain the same number of protons but different numbers of neutrons.

Mass number The sum of the number of neutrons and protons that make up the nucleus of an atom.

Molecule A group of atoms (of the same element or a combination of elements) that are chemically bound together in a fixed ratio.

Nanometer One billionth of a meter.

Nanotechnology The exploration and manipulation of materials whose size is of the order of magnitude of nanometers.

Neutron	One of the basic particles that make up the nucleus of an atom. It is distinguished by having no electric charge.
Nucleus	The central core of an atom, composed of protons and neutrons, that contains all its positive charge and most of its mass.
Oxidation	This term once referred to a chemical reaction in which a substance combined with oxygen but now refers to any reaction in which a substance loses electrons.
Period	A horizontal row in the periodic table.
pH	A measure of the acidity of a solution. A pH of 7 is said to be neutral. The pH decreases as the solution becomes more acidic.
Polymer	A large chain-like molecule made up of repeating smaller molecules that link together.
Radioactive decay	The spontaneous breaking apart of a nucleus of an atom to form a different element. Usually accompanied by the emission of particles and gamma rays.
Reduction	A chemical reaction that once referred to reducing an ore to its pure metal. It now is more generally conceived of as a reaction that involves the gain of electrons.
Salt	A crystalline compound formed from the ions released into solution by an acid or a base.
Superheavy elements	Usually refers to the elements with an atomic number greater than 103.
Transition elements	The elements located between Group IIA and Group IIIA in the periodic table.
Transuranium elements	The elements that follow uranium in the periodic table. These elements are all made artificially.
Valence electrons	The electrons that occupy the outermost shell of an atom. They often determine the chemical behavior of the element.

Some elements—carbon, iron, copper, silver, gold, tin, antimony, mercury, and lead—have been known and used for thousands of years. It is impossible to date their discovery. This chart lists the dates of discovery of all the other elements.

1250
Arsenic discovered by Albertus Magnus.

1669
Phosphorus discovered by Hennig Brand.

1739
Cobalt discovered by Georg Brandt.

1741
Platinum discovered by Charles Wood.

1746
Zinc discovered by Andreas Marggraf.

1751
Nickel isolated by Axel Fredrik Cronstedt.

1753
Bismuth identified by Claude Geoffroy.

1766
Hydrogen discovered by Henry Cavendish.

1772
Nitrogen discovered by Daniel Rutherford.

1774
Chlorine discovered by Carl Wilhelm Scheele; it was identified by Sir Humphry Davy in 1810.

Manganese discovered by Carl Wilhelm Scheele.

Oxygen discovered by Joseph Priestley.

1778
Molybdenum isolated by Carl Wilhelm Scheele.

1782
Tellurium discovered by Franz Joseph von Reichenstein.

1783
Tungsten discovered by Juan Jose and Fausto d'Elhuar y de Suvisa.

1787
Zirconium discovered by Martin Heinrich Klaproth.

1789
Strontium identified by Adair Crawford.

Yttrium identified by Johan Gadolin.

1791
Titanium discovered by Reverend William Gregor.

1797
Chromium discovered by Louis-Nicolas Vauquelin.

1798
Beryllium discovered by Louis-Nicolas Vauquelin.

1801
Vanadium discovered by Andres Manuel del Rio.

Niobium discovered by Charles Hatchett.

1802
Tantalum discovered by Anders Gustav Ekeberg.

1803
Rhodium and **palladium** discovered by William Hyde Wollaston.

Cerium simultaneously discovered by Jöns Jakob Berzelius, Wilhelm Hisinger, and Martin Klaproth.

Osmium and **iridium** discovered by Smithson Tennant.

1807
Sodium and **potassium** isolated by Sir Humphry Davy.

1808
Boron isolated by Sir Humphry Davy, Joseph-Louis Gay-Lussac, and Louis Jacques Thénard.

Magnesium, calcium, and **barium** first isolated and identified by Sir Humphry Davy.

1811
Iodine discovered by Bernard Courtois.

1817
Lithium discovered by Johan August Arfwedson.

Selenium discovered by Jöns Jakob Berzelius.

Cadmium discovered by Friedrich Strohmeyer.

1824
Silicon isolated by Jöns Jakob Berzelius.

1826
Bromine discovered by Antoine-Jérôme Balard.

1827
Aluminum discovered by Hans Christian Oersted.

1828
Thorium discovered by Jöns Jakob Berzelius.

Beryllium discovered by Friedrich Wöhler.

1839
Lanthanum discovered by Carl Gustaf Mosander.

1841
Uranium isolated and identified by Eugène-Melchior Péligot. Henri Becquerel discovered that uranium was radioactive in 1896.

1843
Terbium and **erbium** discovered by Carl Gustaf Mosander.

1844
Ruthenium discovered by K. K. Klaus.

1860
Cesium discovered by Robert Bunsen and Gustav Kirchhoff.

1861
Rubidium discovered by Robert Bunsen and Gustav Kirchhoff.

Thallium discovered by Sir William Crookes.

1863
Indium discovered by Ferdinand Reich.

1868
Helium discovered by Pierre Janssen.

1875
Gallium found and identified by Paul-Émile Lecoq de Boisbaudran.

1878
Ytterbium discovered by Jean de Marignac.

1879
Scandium discovered by Lars Fredrik Nilson.

Samarium discovered by Paul-Émile Lecoq de Boisbaudran.

Holmium and **thulium** discovered by Per Teodor Cleve.

1885
Praseodymium isolated and identified by Carl Auer von Welsbach.

Neodymium discovered by Carl Auer von Welsbach.

1886
Fluorine isolated by Henri Moissan.

Germanium discovered by Clemens Winkler.

Gadolinium discovered by Paul-Émile Lecoq de Boisbaudran and Jean de Marignac.

Dysprosium discovered by Paul-Émile Lecoq de Boisbaudran.

1894
Argon identified by Lord Rayleigh and Sir William Ramsay.

1895
Helium discovered by William Ramsay.

1898
Neon, krypton, and **xenon** discovered by Sir William Ramsay.

Polonium and **radium** discovered by Marie and Pierre Curie.

1899
Actinium discovered by André Debierne.

1900
Radon discovered by Friedrich Ernst Dorn.

1901
Europium isolated by Eugène-Anatole Demarcay.

1907
Lutetium discovered by Carl Auer von Welsbach and Georges Urbain.

1913
Protactinium discovered by Kasimir Fajans and O. H. Gohring.

1923
Hafnium discovered by Dirk Coster and George Karl von Hevesy.

1928
Rhenium discovered by Otto Berg and Wilhelm Noddack.

1937
Technetium discovered by Emilio Segrè and Carlo Perrier.

1939
Francium discovered by Marguerite Perey.

1940
Neptunium first produced by Edwin M. McMillan and Philip H. Abelson.

Astatine created by a team of chemists that included Dale R. Corsun, K. R. Mckenzie, and Emilio Segrè.

1941
Plutonium discovered by Glenn T. Seaborg.

1944
Americium created by a team of scientists led by Glenn T. Seaborg.

Curium created by Glenn T. Seaborg, Ralph A. James, and Albert Ghiorso.

1947
Promethium discovered by J. A. Mirinsky, L. E. Glendenin, and C. D. Coryell.

1949
Berkelium created by Glenn T. Seaborg, Stanley Thompson, and Albert Ghiorso.

1950
Californium created by Stanley Thompson, Kenneth Street Jr., Albert Ghiorso, and Glenn T. Seaborg.

1952
Einsteinium and **fermium** created by a team of scientists led by Albert Ghiorso.

1955
Mendelevium created by a team of scientists led by Albert Ghiorso.

1958
Nobelium identified by a team of scientists led by Albert Ghiorso.

1961
Lawrencium created by a team of scientists that included Albert Ghiorso, T. Sikkeland, A. E. Larsch, and R. M. Latimer.

1969
Rutherfordium created by a team of scientists led by Albert Ghiorso.

1970
Dubnium created by a team of scientists led by Albert Ghiorso.

1981
Bohrium created by a team of scientists led by Peter Armbruster and Gottfried Munzenberg.

1982
Meitnerium created by a team of scientists led by Peter Armbruster and Gottfried Munzenberg.

1984
Hassium created by a team of scientists led by Peter Armbruster and Gottfried Munzenberg.

1994
Darmstadtium and **roentgenium** created by an international team of scientists led by Peter Armbruster.

1996
Copernicium created by an international team of scientists led by Peter Armbruster.

1999
Ununquadium, element 114, created by a collaboration of scientists from the Joint Institute of Nuclear Research in Dubna near Moscow, Russia, and the Lawrence Livermore National Laboratory in California.

2000
Ununhexium created by an international team of scientists working at the Joint Institute for Nuclear Research in Dubna, Russia.

2002
Ununoctium created by the collaboration of an international team of scientists, led by Yuri Oganessian, at the Joint Institute for Nuclear Research in Dubna, Russia.

2004
Ununtrium and **ununpentium** created by a collaboration between Russian scientists and American scientists from the Lawrence Livermore National Laboratory in California, all led by Yuri Oganessian. The work was done at the Joint Institute for Nuclear Research in Dubna, Russia.

2010
Ununseptium created by the collaboration of an international team of scientists led by Yuri Oganessian at the Joint Institute for Nuclear Research in Dubna, Russia.

Further Reading

General Information on Chemistry and the Elements

Atkins, Peter William. *The Periodic Kingdom: A Journey into the Land of the Chemical Elements*. New York: Basic, 1995.

Ball, Philip. *The Elements: A Very Short Introduction.* Oxford: Oxford University Press, 2004.

Cobb, Cathy, and Fetterolf, Monty. *The Joy of Chemistry: The Amazing Science of Familiar Things.* Amherst, N.Y.: Prometheus Books, 2005.

Gray, Theodore, and Mann, Nick: *The Elements: A Visual Exploration of Every Known Atom in the Universe.* New York: Black Dog & Leventhal Publishers, 2009.

Heiserman, David L. *Exploring Chemical Elements and Their Compounds.* Blue Ridge Summit, Pa.: Tab Books, 1992.

Kean, Sam. *The Disappearing Spoon: And Other True Tales of Madness, Love, and the History of the World from the Periodic Table of the Elements.* New York: Little Brown and Company, 2010.

Lewis, Grace Ross. *1001 Chemicals in Everyday Products.* 2d ed. New York: John Wiley, 1998.

Morgan, Nina. *Chemistry in Action: The Molecules of Everyday Life.* New York: Oxford University Press, 1995.

Ratner, Mark A. *Nanotechnology: A Gentle Introduction to the Next Big Idea.* Upper Saddle River, N.J.: Prentice Hall, 2003.

Scerri, Eric R. *The Periodic Table: Its Story and Its Significance.* New York: Oxford University Press, 2006.

Scerri, Eric R. *The Periodic Table: A Very Short Introduction.* Oxford: Oxford University Press, 2011.

Seaborg, Glenn, and Seaborg, Eric. *Adventures in the Atomic Age: From Watt to Washington.* New York: Farrar, Straus and Giroux, 2001.

Snyder, Carl H. *The Extraordinary Chemistry of Ordinary Things.* New York: Wiley, 2002.

Watson, James D. *DNA: The Secret of Life.* New York: Alfred A. Knopf, 2003.

Biographies

Bird, Kai, and Sherwin, Martin. *The Triumph and Tragedy of J. Robert Oppenheimer.* New York: Knopf, Borzoi Books, 2005.

Charles, Daniel. *Master Mind: The Rise and Fall of Fritz Haber, the Nobel Laureate Who Launched the Age of*

Chemical Warfare. New York: HarperCollins, 2005.

Farmelo, Graham. *The Strangest Man: The Hidden Life of Paul Dirac, Mystic of the Atom.* New York: Basic Books, 2009.

Hager, Thomas. *Force of Nature: The Life of Linus Pauling.* New York: Simon & Schuster, 1995.

Krauss, Lawrence. *Quantum Man: Richard Feynman's Life in Science.* New York: W. W. Norton, 2011.

McGrayne, Sharon Bertsch. *Nobel Prize Women in Science: Their Lives, Struggles and Momentous Discoveries.* 2d ed. Washington, D.C.: National Academy Press, 2001.

Pais, Abraham. *Subtle Is the Lord: The Science and the Life of Albert Einstein.* New York: Oxford University Press, 2008.

Pasachoff, Naomi. *Marie Curie and the Science of Radioactivity.* New York: Oxford University Press, 1996.

Quinn, Susan. *Marie Curie: A Life.* New York: Simon & Schuster, 1995.

Strathem, Paul. *Mendeleyev's Dream: The Quest for the Elements.* New York: St. Martin's, 2001.

Wilson, David. *Rutherford: Simple Genius.* Cambridge: Massachusetts Institute of Technology Press, 1983.

www.chemheritage.org
The Chemical Heritage Foundation's website includes digitized oral histories of scientists and a searchable collection of photographs, fine art, artifacts, and scientific instruments from the past 2 centuries. It also presents the history of chemistry in a wide range of articles, based on people and themes.

http://chemicool.com
A color-coded periodic table with a page dedicated to each element that gives the element's basic properties in a series of charts.

www.lbl.gov
The official site of the Berkeley Lab (the Ernest Orlando Lawrence Berkeley National Laboratory). The site contains information about visiting the lab, which offers tours to the public. It is regularly updated with videos and articles about happenings in the scientific community. The site also features links to other sites dedicated to science education.

http://nano.cancer.gov/resource_center/video_journey.asp
Watch videos, listen to podcasts, and read articles about the use of nanotechnology in treating cancer in this website from the National Cancer Institute's Alliance for Nanotechnology in Cancer.

http://nobelprize.org/nobel_prizes/chemistry/laureates
Every Nobel Prize in Chemistry is listed on this website, along with text and videos of the Nobel lectures given by recent winners. The site includes more general articles and videos on chemistry as well.

http://periodic.lanl.gov
A resource for elementary, middle school, and high school students created by the Los Alamos National Laboratory's Chemistry Division. The site provides an interactive periodic table with links to a description of each element.

www.webelements.com
This site features an interactive periodic table with links to a page for each element. Each page has summary information about the element and further links giving such details as electronic, physical, and nuclear properties; crystallography; and a proper compound index. The site also contains a number of useful graphs and other graphics.

Index

References to main entries are indicated by **bold** page numbers; references to illustrations are indicated by page numbers in *italics*.

Picture Credits

Frontispiece (from top, left to right): Private collection, Private collection, National Park Service, Photo by Peter Jones, Library of Congress, LC-DIG-matpc-02659, Private collection, Private collection, Private collection; 8 (top): AIP Emilio Segrè Visual Archives; 8 (bottom): New York Public Library, General Research Division, Astor, Lenox, and Tilden Foundations; 17: AIP Emilio Segrè Visual Archives, Brittle Books Collection; 18: NASA/Photo KSC-2011-3753 by Sandra Joseph and Kevin O'Connell; 20: Library of Congress, POS 6-U.S., no. 1176 (C size) [P&P]; 23: McGill University, Department of Physics; 24: Courtesy Goodyear Tire and Rubber Company; 25: Photo Researchers, Inc., photo by Charles Winter; 27: Lawrence Berkeley National Laboratory, courtesy AIP Emilio Segrè Visual Archives; 29: Smithsonian Institution, photo by Dane A. Penland; 30: AIP Emilio Segrè Visual Archives, W. F. Meggers Gallery of Nobel Laureates; 31: National Bureau of Standards Archives, courtesy AIP Emilio Segrè Visual Archives; 32: Courtesy U.S. Borax; 35: Photo Researchers, Inc., photo by Charles Winter; 36: Shutterstock, 64111444; 39: Princeton Plasma Physics Laboratory; 42: Private collection; 43: Courtesy Delphi Automotive Systems; 45: Shutterstock, 59885752; 47: Private collection; 49: Private collection; 51: Colgate–Palmolive; 53: Las Vegas News Bureau; 55: Arm and Hammer; 56: Photo Researchers, Inc., Photo by Charles Winter; 58: Courtesy Bayer Corporation; 60: Photo Researchers, Inc., photo by Charles Winter; 61: AT&T Archives; 63: AT&T Archives; 65: Library of Congress, LC-USZ62-40314; 68: National Park Service, Yellowstone National Park; 69: National Archives, NA-052-S-2303; 71: AIP Emilio Segrè Visual Archives, *Physics Today* Collection; 72: Private collection; 74: Shutterstock, 71425288; 76: Photo Researchers, Inc., photo by Charles Winter; 77: National Park Service, Photo by Peter Jones; 81: Private collection; 82: Courtesy Pratt and Whitney; 85: Smithsonian Institution, photo by Dane A. Penland; 90: Courtesy Bethlehem Steel, 77108A-4; 93: Private collection; 94: Image No. 31516, American Museum of Natural History Library; 95: U.S. Mint; 96: Courtesy Olin Corporation; 98: Private collection; 99: U.S. Mint; 100: Shutterstock, 1090883; 101: Shutterstock, 62486296; 102: Photo Researchers, Inc., photo by Charles Winter; 105: Museum of Modern Art, Film Stills Archive; 106: Shutterstock, 22080349; 109: Courtesy Eastman Kodak; 110: AIP Emilio Segrè Visual Archives, W. F. Meggers Gallery of Nobel Laureates; 114: Fireworks by Grucci, Inc.; 115: NASA; 118: Private collection; 120: Nuclear Energy Institute; 127: Library of Congress; 128: Courtesy Oneida Silversmiths; 131: © Robert Visser, Greenpeace; 133: Metropolitan Museum of Art, Gift of Joseph Francis, 1943 (43.162.29); 135: Private collection; 138: Private collection; 139: Courtesy Eastman Kodak; 141: Argonne National Laboratory; 142: Photographische Gesellschaft, Berlin, courtesy AIP Emilio Segrè Visual Archives, W. F. Meggers Collection, Brittle Books Collection, Harvard University Collection; 143: Courtesy of Bureau International des Poids et Mesures; 145: Photo Researchers, Inc., © CNRI/SPL/Science Source; 148: Courtesy of Maytag, Whirlpool Digital Library; 153: Oak Ridge National Laboratory; 156: Private collection; 171: Courtesy GE Lighting; 173: Courtesy Parker Pen USA; 175: Courtesy of Bureau International des Poids et Mesures; 177: Courtesy of Bureau International des Poids et Mesures; 178: California State Library, California Section; 179: Courtesy Santa Fe Pacific Gold Corporation; 180: Shutterstock, 33172321; 187: Photo Researchers, Inc., photo by Charles Winter; 189: Argonne National Laboratory; 191: Archives Curie et Joliot-Curie; 193: Lawrence Berkeley National Laboratory, courtesy AIP Emilio Segrè Visual Archives; 195: Courtesy BRK Brands, Inc.; 197: AIP Emilio Segrè Visual Archives; 199: Argonne National Laboratory; 201: Courtesy Suunto USA; 204: University of Chicago Library, Department of Speical Collections; 205: University of Pittsburgh, Hillman Library; 206: *Oakland Tribune,* courtesy AIP Emilio Segrè Visual Archives, Fermi Film Collection; 207: NASA; 209: Argonne National Laboratory, courtesy AIP Emilio Segrè Visual Archives; 211: Lawrence Berkeley National Laboratory, courtesy AIP Emilio Segrè Visual Archives; 213: Lawrence Berkeley National Laboratory; 214: Oak Ridge National Laboratory; 218: Lawrence Berkeley National Laboratory, courtesy AIP Emilio Segrè Visual Archives; 220: Leningrad Physico-Technical Institute, courtesy AIP Emilio Segrè Visual Archives; 225: Courtesy Peter Armbruster; 227: Photo courtesy of Lawrence Berkeley National Laboratory; 232: Lawrence Livermore National Laboratory; 236: Photo Researchers on behalf of Science Photo Library; 238: GSI Helmholtzzentrum für Schwerionenforschung GmbH, photo by J. Mai

About the Author

Albert Stwertka is Professor Emeritus at the United States Merchant Marine Academy and previously served as the head of the mathematics and science department there. Dr. Stwertka conducted research in the field of atomic physics and was instrumental in establishing training programs for the N.S. *Savannah* nuclear ship during the Atoms for Peace program. He is the author of numerous books on science and math, including *The World of Atoms and Quarks, Recent Revolutions in Physics, Recent Revolutions in Mathematics,* and *Physics: From Newton to the Big Bang.*